SOLUTIONS AND PROBLEM-SOLVING MANUAL
FOR

Genetics Essentials:
Concepts and Connections

Fourth Edition

JUNG H. CHOI
MARK E. MCCALLUM

W. H. FREEMAN AND COMPANY
NEW YORK

ISBN-10: 1-319-20061-3
ISBN-13: 978-1-319-20061-9

Printed in the United States of America

First printing

W. H. Freeman and Company
One New York Plaza
Suite 4500
New York, NY 10004-1562

www.macmillanlearning.com

Contents

Chapter One: Introduction to Genetics

COMPREHENSION QUESTIONS

Section 1.1

*1. How did the Hopi culture contribute to the high incidence of albinism among members of the Hopi tribe?

Solution:
In the Hopi culture, people with albinism were considered special and awarded special status in the village. Hopi male albinos were not required to work the fields, thus avoiding extensive exposure to sunlight that could prove damaging or deadly. Albinism was considered a positive trait and not a negative physical condition, which allowed albinos to have more children, increasing the frequency of the albino allele. Finally, the small population size of the Hopi tribe may have helped increase the allele frequency of the albino gene due to chance.

2. Give at least three examples of the role of genetics in society today.

Solution:
Genetics plays important roles in the diagnosis and treatment of hereditary diseases, in breeding plants and animals for improved production and disease resistance, and in producing pharmaceuticals and novel crops through genetic engineering.

3. Briefly explain why genetics is crucial to modern biology.

Solution:
Genetics is crucial to modern biology in that it provides unifying principles: all organisms use nucleic acid as their genetic material, and all organisms encode genetic information in the same manner. The study of many other biological disciplines, such as developmental biology, ecology, and evolutionary biology, is supported by genetics.

4. List the three traditional subdisciplines of genetics and summarize what each covers.

Solution:
Transmission genetics: inheritance of genes from one generation to the next, gene-mapping.
Molecular genetics: structure, organization, and function of genes at a molecular level.
Population genetics: the genetic composition of populations and how the genetic composition changes over time.

5.   What are some characteristics of model genetic organisms that make them useful for genetic studies?

Solution:
Model genetic organisms have relatively short generation times, produce numerous progeny, are amenable to laboratory manipulations, and can be maintained and propagated inexpensively.

Section 1.2

6.   When and where did agriculture first arise? What role did genetics play in the development of the first domesticated plants and animals?

Solution:
Agriculture first arose 10,000 to 12,000 years ago in the area now referred to as the Middle East (i.e., Turkey, Iran, Iraq, Syria, Jordan, and Israel). Early farmers selectively bred individual wild plants or animals that had useful characteristics with others that had similar useful traits. The farmers then selected for offspring that contained those useful features. Early farmers did not completely understand genetics, but they clearly understood that breeding individual plants or animals with desirable traits would lead to offspring that contained these same traits. This selective breeding led to the development of domesticated plants and animals.

7.   Outline the concept of pangenesis and explain how it differs from the germ-plasm theory.

Solution:
Pangenesis theorizes that information for creating each part of the offspring's body originates from each part of the parent's body and is passed through the reproductive organs to the embryo at conception. Pangenesis suggests that changes in parts of the parent's body may be passed to the offspring's body. The germ-plasm theory, in contrast, states that the reproductive cells possess all of the information required to make the complete body; the rest of the body contributes no information to the next generation.

8.   What does the concept of the inheritance of acquired characteristics propose and how is it related to the notion of pangenesis?

Solution:
The theory of inheritance of acquired characteristics postulates that traits acquired during one's lifetime can be transmitted to offspring. It developed from pangenesis, which postulates that information from all parts of one's body is transmitted to the next generation. Thus, for example, learning acquired in the brain or larger arm muscles developed through exercise could be transmitted to offspring.

9.   What is preformationism? What did it have to say about how traits are inherited?

Solution:
Preformationism is the theory that the offspring results from a miniature adult form already preformed in the sperm or the egg. All traits would thus be inherited from only one parent, either the father or the mother, depending on whether the homunculus (the preformed miniature adult) resided in the sperm or the egg.

10.   Define blending inheritance and contrast it with preformationism.

Solution:
The theory of blending inheritance proposes that the egg and sperm from two parents contain material that blends upon conception, influencing the development of the offspring. This theory indicates that the offspring is an equal blend of the two parents. In preformationism, the offspring inherits all of its traits from one parent.

11.   How did developments in botany during the seventeenth and eighteenth centuries contribute to the rise of modern genetics?

Solution:
Botanists of the seventeenth and eighteenth centuries developed new techniques for crossing plants and creating plant hybrids. These early experiments provided essential background work for Mendel's plant crosses. Mendel's work laid the foundation for the study of modern genetics.

12.   List some advances in genetics made in the twentieth century.

Solution:
1902	Proposal that genes are located on chromosomes by Walter Sutton
1910	Discovery of the first genetic mutation in a fruit fly by Thomas Hunt Morgan
1930	The foundation of population genetics by Ronald A. Fisher, John B. S. Haldane, and Sewall Wright
1940s	The use of viral and bacterial genetic systems
1953	Three-dimensional structure of DNA described by Watson and Crick
1966	Deciphering of the genetic code
1973	Recombinant DNA experiments
1977	Chemical and enzymatic methods for DNA sequencing developed by Walter Gilbert and Frederick Sanger
1986	PCR developed by Kary Mullis
1990	Gene therapy

13.   Briefly explain the contribution that each of the following persons made to the study of genetics.

Solution:

a.   Matthias Schleiden and Theodor Schwann

Proposed the concept of the cell theory, which indicated that the cell is the fundamental unit of living organisms. Caused biologists interested in heredity to examine cell reproduction.

b.   August Weismann

Proposed the germ-plasm theory, which holds that cells in reproductive organs carry a complete set of genetic information.

c.   Gregor Mendel

First discovered the basic rules of inheritance.

d.   James Watson and Francis Crick

Along with Rosalind Franklin and Maurice Wilkins, described the three-dimensional structure of DNA.

e.   Kary Mullis

Developed the polymerase chain reaction, which is used to quickly amplify small amounts of DNA.

Section 1.3

14.   What are the two basic cell types (from a structural perspective) and how do they differ?

Solution:
The two basic cell types are prokaryotic and eukaryotic. Prokaryotic cells do not have a nucleus, and their chromosomes are found within the cytoplasm. They do not possess membrane-bound cell organelles. Eukaryotic cells possess a nucleus and membrane-bound cell organelles.

*15.   Summarize the relations between genes, DNA, and chromosomes.

Solution:
Genes are composed of DNA nucleotide sequences that are located at specific positions in chromosomes.

APPLICATION QUESTIONS AND PROBLEMS

Section 1.1

*16.   How are genetics and evolution related?

Solution:
Evolution is genetic change over time. For evolution to occur, genetic variation must first arise, and then evolutionary forces change the proportion of genetic variants over time. Genetic variation is therefore the basis of all evolutionary change.

*17.   For each of the following genetic topics, indicate whether it focuses on transmission genetics, molecular genetics, or population genetics.

a.   Analysis of pedigrees to determine the probability of someone inheriting a trait

Solution: Transmission genetics

b.   Study of people on a small island to determine why a genetic form of asthma is so prevalent on the island

Solution: Population genetics

c.   Effect of nonrandom mating on the distribution of genotypes among a group of animals

Solution: Population genetics

d.   Examination of the nucleotide sequences found at the ends of chromosomes

Solution: Molecular genetics

e.   Mechanisms that ensure a high degree of accuracy during DNA replication

Solution: Molecular genetics

f.   Study of how the inheritance of traits encoded by genes on sex chromosomes (sex-linked traits) differs from the inheritance of traits encoded by genes on nonsex chromosomes (autosomal traits)

Solution: Transmission genetics

Section 1.2

*18. Genetics is said to be both a very old science and a very young science. Explain what this means.

Solution:
Genetics is old in the sense that humans have been aware of hereditary principles for thousands of years and have applied these principles since the beginning of agriculture. It is very young in the sense that the fundamental principles were not uncovered until Mendel's time, and the discovery of the structure of DNA and the principles of recombinant DNA have occurred within the last 60 years.

*19. Match each description (*a* through *d*) with the correct theory or concept listed below.

Preformationism

Pangenesis

Germ-plasm theory

Inheritance of acquired characteristics

a. Each reproductive cell contains a complete set of genetic information.

Solution: Germ-plasm theory

b. All traits are inherited from one parent.

Solution: Preformationism

c. Genetic information may be altered by the use of a characteristic.

Solution: Inheritance of acquired characteristics

d. Cells of different tissues contain different genetic information.

Solution: Pangenesis

*20. Compare and contrast the following ideas about inheritance.

a. Pangenesis and germ-plasm theory

Solution:
Pangenesis postulates that particles carry genetic information from all parts of the body to the reproductive organs, and then the genetic information is conveyed to the embryo, where each unit directs the formation of its own specific part of the body. According to the germ-plasm theory, gamete-producing cells found within the reproductive organs contain the complete set of genetic information that is passed to the gametes. Pangenesis and the germ-plasm theory are similar in that both propose

that genetic information is contained in discrete units that are passed on to the offspring. They differ in where that genetic information resides. In pangenesis, it resides in different parts of the body and must travel to the reproductive organs. In the germ-plasm theory, all the genetic information is already in the reproductive cells.

b. Preformationism and blending inheritance

Solution:
Preformationism holds that the sperm or egg contains a miniature preformed adult called the homunculus. In development, the homunculus grows to produce an offspring. Only one parent contributes genetic traits to the offspring. Blending inheritance requires contributions of genetic material from both parents. The genetic contributions from the parents blend to produce the genetic material of the offspring. Having been blended, the genetic material cannot be separated for future generations.

c. The inheritance of acquired characteristics and our modern theory of heredity

Solution:
The inheritance of acquired characteristics postulates that traits acquired in a person's lifetime alter the genetic material and can be transmitted to offspring. Our modern theory of heredity indicates that offspring inherit genes located on chromosomes passed from their parents. These chromosomes segregate in meiosis in the parent's germ cells and are passed into the gametes.

Section 1.3

*21. Compare and contrast the following terms:

a. Eukaryotic and prokaryotic cells

Solution:
Both cell types have lipid bilayer membranes, DNA genomes, and machinery for DNA replication, transcription, translation, energy metabolism, response to stimuli, growth, and reproduction. Eukaryotic cells have a nucleus containing chromosomal DNA and possess internal membrane-bound organelles.

b. Gene and allele

Solution:
A gene is a basic unit of hereditary information, usually encoding a functional RNA or polypeptide. Alleles are variant forms of a gene, arising through mutation.

c. Genotype and phenotype

Solution:
The genotype is the set of genes or alleles inherited by an organism from its parent(s). The expression of the genes of a particular genotype, through interaction with environmental factors, produces the phenotype, the observable trait.

d. DNA and RNA

Solution:
Both are nucleic acid polymers. RNA contains a ribose sugar, whereas DNA contains deoxyribose sugar. RNA also contains uracil as one of the four bases, whereas DNA contains thymine. The other three bases are common to both DNA and RNA. Finally, DNA is usually double-stranded, consisting of two complementary strands, whereas RNA is single-stranded.

e. DNA and chromosome

Solution:
Chromosomes are structures consisting of DNA and associated proteins. The DNA contains the genetic information.

CHALLENGE QUESTIONS

Introduction

*22. The type of albinism that arises with high frequency among the Hopis (discussed in the introduction to this chapter) is most likely oculocutaneous albinism type 2, which is caused by a defect in the *OCA2* gene on chromosome 15. Do some research on the Internet to determine how the phenotype of this type of albinism differs from phenotypes of other forms of albinism in humans and the mutated genes that result in these phenotypes. Hint: Visit the website Online Mendelian Inheritance in Man and search the database for albinism.

Solution:

Type of albinism	Phenotype	Gene mutated
OCA2	Pigment reduced in skin, hair, and eyes, but small amount of pigment acquired with age; visual problems	*OCA2*
OCA1B	General absence of pigment in hair, skin, and eyes, but there may be small amount of pigment; does not vary with age; visual problems	Gene that encodes tyrosinase
OCA1A	Complete absence of pigment; visual problems	Gene that encodes tyrosinase
OCA3	Some pigment present; sun sensitivity and visual problems	Gene that encodes tyrosinase-related protein 1
OASD	Lack of pigment in the eyes and deafness later in life	Unknown
OA1	Lack of pigment in the eyes but normal elsewhere	*GPR143*
ROCA	Bright copper-red coloration in skin and hair of Africans; dilution of color in iris	Gene that encodes tyrosinase-related protein 1
OCA4	Reduced pigmentation	*MATP*

Section 1.1

23.   We now know a great deal about the genetics of humans. What are some of the reasons humans have been the focus of intensive genetic study?

Solution:
Humans are intensively interested in how humans function biologically. Because of this intense interest, we know more about the anatomy, physiology, genetics, and biochemistry of humans than of many other organisms. Many human diseases and disorders are associated with human genes. Understanding how to treat and diagnose these diseases and disorders requires intensive studies to identify the gene(s) responsible for the disorder as well as understanding how they are inherited and expressed. Recent advances in the understanding of genetic risk factors associated with diseases such as heart disease and cancer have enabled the development of predictive genetic tests for some of these disorders. These successes continue to stimulate a focus in identifying genetic risk factors for other diseases. The ability of families to keep careful records about members extending back many generations has facilitated the study of human inheritance that has aided the ability of researchers to identify genetic markers within families. In addition, these detailed records have provided some humans who are intensely interested in their own heredity the ability to trace their ancestry.

24.   Describe some of the ways in which your own genetic makeup affects you as a person. Be as specific as you can.

Solution:
Answers will vary but should include observations similar to those in the following example: Genes affect my physical appearance; for example, they probably have largely determined the fact that I have brown hair and brown eyes. Undoubtedly, genes have affected my height of five feet, seven inches, which is quite close to the height of my father and mother, and my slim build. My dark complexion mirrors the skin color of my mother. I have inherited susceptibilities to certain diseases and disorders that tend to run in my family; these include asthma, a slight tremor of the hand, and vertigo.

25.   Describe at least one trait that appears to run in your family (appears in multiple members of the family). Do you think that this trait runs in your family because it is an inherited trait or because it is caused by environmental factors that are common to family members? How might you distinguish between these possibilities?

Solution:
My two brothers and I share two traits: all three of us are both taciturn (we don't speak much) and smart (just don't ask my teenage daughter). Although the literature provides evidence for a genetic component for intelligence, I'm not aware of any studies on the heritability of being taciturn. If I were to investigate the extent to which these traits are determined by the environment or by heredity, I would look at studies of twins who had been separated at birth and lived in different environments to adulthood. Such studies would separate environmental factors from genetic factors, whereas studies of family

members reared in the same household are confounded by the fact that the family members experienced similar environments. If the trait had a strong genetic component, we would expect identical twins reared apart to be similarly taciturn or similarly intelligent. One would have to devise some objective measure of these traits—degrees of being taciturn or smart.

Section 1.3

*26.   Suppose that life exists elsewhere in the universe. All life must contain some type of genetic information, but alien genomes might not consist of nucleic acids and have the same features as those found in the genomes of life on Earth. What do you think might be the common features of all genomes, no matter where they exist?

Solution:
All genomes must have the ability to store complex information and to vary. The blueprint for the entire organism must be contained within the genome of each reproductive cell. The information has to be in the form of a code that can be used as a set of instructions for assembling the components of the cells. The genetic material of any organism must be stable, be replicated precisely, and be transmitted faithfully to the progeny but must be capable of mutating.

27.   Choose one of the ethical or social issues in parts *a* through *d* and give your opinion on the issue. For background information, you might read one of the articles on ethics listed and marked with an asterisk in the Suggested Readings section for Chapter 1 in your SaplingPlus.

a.   Should a person's genetic makeup be used in determining his or her eligibility for life insurance?

Solution:
Arguments pro: Genetic susceptibility to certain types of diseases or conditions is relevant information regarding consequences of exposure to certain occupational hazards. Genes that will result in neurodegenerative diseases, such as Huntington disease, Alzheimer disease, or breast cancer, could logically be considered preexisting conditions. Insurance companies have a right, and arguably a duty to their customers, to exclude people with genetic preconditions so that insurance rates can be lowered for the general population.
Arguments con: The whole idea of insurance is to spread the risk and pool assets. Excluding people based on their genetic makeup would deny insurance to people who need it most. Indeed, as information about various genetic risks accumulates, more people would become excluded until only a small fraction of the population is insurable.

b. Should biotechnology companies be able to patent newly sequenced genes?

Solution:
Pro: Patenting genes provides companies with protection for their investment in research and development of new drugs and therapies. Without such patent protection, companies would have less incentive to expend large amounts of money in genetic research, which would slow the pace of advancement of medical research. Such a result would be detrimental to everyone.
Con: Patents on human genes would be like allowing companies to patent a human arm. Genes are integral parts of our selves, so how can a company patent something that every human has?

c. Should gene therapy be used on people?

Solution:
Pro: Gene therapy can be used to cure previously incurable or intractable genetic disorders and to relieve the suffering of millions of people.
Con: Gene therapy may lead to genetic engineering of people for unsavory ends. Who determines what is a genetic defect? For example, would short stature be considered a genetic defect?

d. Should genetic testing for inherited disorders for which there is no treatment or cure be made available?

Solution:
Pro: Information will provide relief from unnecessary anxiety (if the test is negative). Even if the test result is positive for a genetic disorder, it provides the person, the family, and friends with information and time to prepare. Information about one's own genetic makeup is a right; every person should be able to make his or her own choice as to whether he or she wants this information.
Con: If there is no treatment or cure, a positive test result can have no good consequences. Receiving such a result would be like receiving a death sentence or a sentence of extended punishment. It will only engender feelings of hopelessness and depression and may cause some people to terminate their own lives prematurely.

*28. A 45-year-old woman undergoes genetic testing and discovers that she is at high risk for developing colon cancer and Alzheimer disease. Because her children have 50% of her genes, they also may be at an increased risk for these diseases. Does she have a moral or legal obligation to tell her children and other close relatives about the results of her genetic testing?

Solution:
Legally, she is not required to inform her children or other relatives about her test results, but people may have different opinions about her moral and parental responsibilities. On the one hand, she has the legal right to keep private the results of any medical information, including the results of genetic testing. On the other hand, her children may be at an

increased risk of developing these disorders and might benefit from that knowledge. For example, the risk of colon cancer can be reduced by regular examinations so that tumors can be detected and removed before they become cancerous. Some people might argue that her parental responsibilities include providing her children with information about possible medical problems. Another issue to consider is the possibility that her children or other relatives might not want to know their genetic risk, particularly for a disorder such as Alzheimer disease for which there is no cure.

*29.   Suppose that you could undergo genetic testing at age 18 for susceptibility to a genetic disease that would not appear until middle age and has no available treatment.

    a.   What would be some of the possible reasons for having or not having such a genetic test?

        Solution:
        Having the genetic test removes doubt about the potential for the disorder: you are either susceptible or not. Knowing about the potential of a genetic disorder enables you to make lifestyle changes that might lessen the effect of the disease or lessen the risk. The types and nature of future medical tests could be positively affected by the genetic testing, thus allowing for early warning and screening for the disease. The knowledge could also enable you to make informed decisions about whether to have children and the potential of passing the trait to your offspring. Additionally, by knowing what to expect, you could plan your life accordingly.

        Reasons for not having the test typically concern the potential for testing positive for the susceptibility to the genetic disease. If the susceptibility is detected, there is potential for discrimination. For example, your employer (or possibly a future employer) might consider you a long-term liability, thus affecting employment options. Insurance companies may not want to insure you for the condition or its symptoms, and social stigmatism regarding the disease could be a factor. Knowledge of the potential future condition could lead to psychological difficulties in coping with the anxiety of waiting for the disease to manifest.

    b.   Would you personally want to be tested? Explain your reasoning.

        Solution:
        There is no "correct" answer, but some of the reasons for wanting to be tested are as follows: The test would remove doubt about the susceptibility, particularly if family members have had the genetic disease; either a positive or negative result would allow for informed planning of lifestyle, medical testing, and family choices in the future.

Chapter Two: Chromosomes and Cellular Reproduction

COMPREHENSION QUESTIONS

Section 2.1

1. What are some genetic differences between prokaryotic and eukaryotic cells?

    Solution:

Prokaryotic cell	Eukaryotic cell
No nucleus	Nucleus present
No paired chromosomes (haploid)	Paired chromosomes common (diploid)
Typically single circular chromosome containing a single origin of replication	Typically multiple linear chromosomes containing centromeres, telomeres, and multiple origins of replication
Single chromosome is replicated with each copy moving to opposite sides of the cell	Chromosomes are replicated and segregate during mitosis or meiosis to the proper location
No histone proteins bound to DNA	Histone proteins are bound to DNA

2. Why are viruses often used in the study of genetics?

    Solution:
    Because of their simplicity and small genome size.

Section 2.2

3. List three fundamental events that must take place in cell reproduction.

    Solution:
    The three events are (1) a cell's genetic information must be copied, (2) the copies of the genetic information must be separated from one another, and (3) the cell must divide.

4. Name three essential structural elements of a functional eukaryotic chromosome and describe their functions.

    Solution:
    (1) Centromere: serves as the point of attachment for the spindle microtubules.
    (2) Telomeres, or the natural ends of the linear eukaryotic chromosome: serve to stabilize the ends of the chromosome; may have a role in limiting cell division.
    (3) Origins of replication: serve as the starting place for DNA synthesis.

5.   Sketch and identify four different types of chromosomes based on the position of the centromere.

Solution:

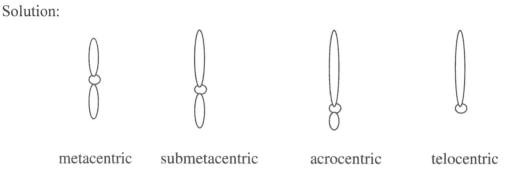

metacentric   submetacentric   acrocentric   telocentric

6.   List the stages of interphase and the major events that take place in each stage.

Solution:
Three predominant stages are found in interphase of cells active in the cell cycle.
(1) **G_1 (Gap 1).** In this phase, the cell grows and synthesizes proteins necessary for cell division. During G_1, the G_1/S checkpoint takes place. Once the cell has passed this checkpoint, it is committed to divide.
(2) **S phase.** During S phase, DNA replication takes place.
(3) **G_2 (Gap 2).** In G_2, additional biochemical reactions take place that prepare the cell for mitosis. A major checkpoint in G_2 is the G_2/M checkpoint. Once the cell has passed this checkpoint, it enters into mitosis.
A fourth stage is frequently found in cells prior to the G_1/S checkpoint. Cells may exit the active cell cycle and enter into a nondividing stage called G_0.

7.   What are checkpoints? List some of the important checkpoints in the cell cycle.

Solution:
Checkpoints function to ensure that all the cellular components, such as important proteins and chromosomes, are present and functioning before the cell moves to the next stage of the cell cycle. If components are missing or not functioning, the checkpoint will prevent the cell from moving to the next stage. The checkpoints prevent defective cells from replicating and malfunctioning.

These checkpoints occur throughout the various stages of the cell cycle. Important checkpoints include the G_1/S checkpoint, which occurs during G_1 prior to the S phase; the G_2/M checkpoint, which occurs in G_2 prior to mitosis; and the spindle-assembly checkpoint, which occurs during mitosis.

8.   List the stages of mitosis and the major events that take place in each stage.

Solution:
(1) **Prophase:** The chromosomes condense and become visible, the centrosomes move apart, and microtubule fibers form from the centrosomes.
(2) **Prometaphase:** The nucleoli disappear and the nuclear envelope begins to disintegrate, allowing for the cytoplasm and nucleoplasm to join. The sister chromatids of each chromosome are attached to microtubules from the opposite centrosomes.
(3) **Metaphase:** The spindle microtubules are clearly visible and the chromosomes arrange themselves on the equatorial plane of the cell.
(4) **Anaphase:** The sister chromatids separate at the centromeres after the breakdown of cohesin protein, and the newly formed daughter chromosomes move to the opposite poles of the cell.
(5) **Telophase:** The nuclear envelope reforms around each set of daughter chromosomes. Nucleoli reappear. Spindle microtubules disintegrate.

9.   What are the genetically important results of the cell cycle and mitosis?

Solution:
In the cell cycle, the genetic material is precisely copied and mitosis ensures that the identical copies of the genetic material are separated accurately into the new daughter cells, resulting in two cells containing the same genetic information. In other words, the cells have genomes identical to each other and to the mother cell.

10.   Why are the two cells produced by the cell cycle genetically identical?

Solution:
The two cells are genetically identical because during the S phase an exact copy of each DNA molecule was created. These exact copies give rise to the two identical sister chromatids. Mitosis ensures that each new cell receives one copy of the two identical sister chromatids. Thus, the newly formed cells will contain identical daughter chromosomes.

Section 2.3

11.   What are the stages of meiosis and what major events take place in each stage?

Solution:
Meiosis I: Separation of homologous chromosomes

Prophase I: The chromosomes condense and homologous pairs of chromosomes undergo synapsis. While the chromosomes are synapsed, crossing over occurs. The nuclear membrane disintegrates and the meiotic spindle begins to form.

Metaphase I: The homologous pairs of chromosomes line up on the equatorial plane of the metaphase plate.

Anaphase I: Homologous chromosomes separate and move to opposite poles of the cell. Each chromosome possesses two sister chromatids.

Telophase I: The separated homologous chromosomes reach the spindle poles and are at opposite ends of the cell.

Meiosis I is followed by cytokinesis, resulting in the division of the cytoplasm and the production of two haploid cells. These cells may skip directly into meiosis II or enter interkinesis, where the nuclear envelope reforms and the spindle fibers break down.

Meiosis II: Separation of sister chromatids

Prophase II: Chromosomes condense, the nuclear envelope breaks down, and the spindle fibers form.

Metaphase II: Chromosomes line up at the equatorial plane of the metaphase plate.

Anaphase II: The centromeres split, which results in the separation of sister chromatids.

Telophase II: The daughter chromosomes arrive at the poles of the spindle. The nuclear envelope reforms, and the spindle fibers break down.
Following meiosis II, cytokinesis takes place.

12.   What are the major results of meiosis?

Solution:
Meiosis comprises two cell divisions, thus producing four new cells (in many species). The chromosome number of a haploid cell produced by meiosis I (haploid) is half the chromosome number of the original diploid cell. Finally, the cells produced by meiosis are genetically different from the original cell and genetically different from one another.

13.   What two processes unique to meiosis are responsible for genetic variation? At what point in meiosis do these processes take place?

Solution:
(1) Crossing over, which begins during the zygotene stage of prophase I and is completed near the end of prophase I.
(2) The random distribution of chromosomes to the daughter cells, which takes place in anaphase I of meiosis.

14.   List some similarities and differences between mitosis and meiosis. Which differences do you think are most important and why?

Solution:

Mitosis	Meiosis
A single cell division produces two genetically identical progeny cells.	Two cell divisions usually result in four progeny cells that are not genetically identical.
Chromosome number of progeny cells and the original cell remain the same.	Daughter cells are haploid and have half the chromosomal complement of the original diploid cell as a result of the separation of homologous pairs during anaphase I.
Daughter cells and the original cell are genetically identical. No separation of homologous chromosomes or crossing over takes place.	Crossing over in prophase I and separation of homologous pairs in anaphase I produce daughter cells that are genetically different from one another and from the original cell.
Homologous chromosomes do not synapse.	Synapsis of homologous chromosomes takes place in prophase I.
In metaphase, individual chromosomes line up on the metaphase plate.	In metaphase I, homologous pairs of chromosomes line up on the metaphase plate. Individual chromosomes line up in metaphase II.
In anaphase, sister chromatids separate.	In anaphase I, homologous chromosomes separate. Separation of sister chromatids takes place in anaphase II.

A key difference is that mitosis produces cells genetically identical with each other and with the original cell, resulting in the orderly passage of information from one cell to its progeny. In contrast, by producing progeny that do not contain pairs of homologous chromosomes, meiosis results in the reduction of chromosome number from that of the original cell. Meiosis also allows for genetic variation through crossing over and the random assortment of homologous chromosomes.

15.   Outline the processes of spermatogenesis and oogenesis in animals.

Solution:
In animals, spermatogenesis occurs in the testes. Primordial diploid germ cells divide mitotically to produce diploid spermatogonia that can either divide repeatedly by mitosis or enter meiosis. A spermatogonium that has entered prophase I of meiosis is called a primary spermatocyte and is diploid. Upon completion of meiosis I, two haploid cells, called secondary spermatocytes, are produced. Upon completing meiosis II, the secondary spermatocytes produce a total of four haploid spermatids.

Female animals produce eggs through the process of oogenesis. Similar to what takes place in spermatogenesis, primordial diploid cells divide mitotically to produce diploid oogonia that can divide repeatedly by mitosis, or enter meiosis. An oogonium that has entered prophase I is called a primary oocyte and is diploid. Upon completion of meiosis I, the cell divides, but unequally. One of the newly produced haploid cells receives most of the cytoplasm and is called the secondary oocyte. The other haploid cell receives only a small portion of the cytoplasm and is called the first polar body. Ultimately, the secondary oocyte will complete meiosis II and produce two haploid cells. One cell, the ovum, will receive most of the cytoplasm from the secondary oocyte. The smaller haploid cell is called the second polar body. Typically, the polar bodies disintegrate, and only the ovum is capable of being fertilized.

16.   Outline the processes of male and female gamete formation in plants.

Solution:
Plants alternate between a multicellular haploid stage called the gametophyte and a multicellular diploid stage called the sporophyte. Meiosis in the diploid sporophyte stage of plants produces haploid spores that develop into the gametophyte. The gametophyte produces gametes by mitosis.

In flowering plants, the microsporocytes found in the stamen of the flower undergo meiosis to produce four haploid microspores. Each microspore divides by mitosis to produce the pollen grain, (or the microgametophyte). Within the pollen grain are two haploid nuclei. One of the haploid nuclei divides by mitosis to produce two sperm cells. The other haploid nucleus directs the formation of the pollen tube.

Female gamete production in flowering plants takes place within the megagametophyte. Megasporocytes found within the ovary of a flower divide by meiosis to produce four megaspores. Three of the megaspores disintegrate, while the remaining megaspore divides mitotically to produce eight nuclei that form the embryo sac (or female gametophyte). Of the eight nuclei, one will become the egg.

APPLICATION QUESTIONS AND PROBLEMS

Introduction

17.   Answer the following questions for the Blind Men's Riddle, presented at the beginning of the chapter.

a. What component of the cell cycle do the two socks of a pair represent?

Solution:
The two chromatids of a chromosome

b. In the riddle, each blind man buys his own pairs of socks, but the clerk places all pairs into one bag. Thus, there are two pairs of socks of each color in the bag (two black pairs, two blue pairs, two gray pairs, etc.). What do the two pairs (four socks in all) of each color represent?

Solution:
The two chromosomes of a homologous pair

c. What in the riddle performs the same function as spindle fibers?

Solution:
The hands of the two blind men

Section 2.1

18. A cell has a circular chromosome and no nuclear membrane. Its DNA is associated with some histone proteins. Does this cell belong to the bacteria, the archaea, or the eukaryotes? Explain your reasoning.

Solution:
This cell is most likely an archaea. The cell is not eukaryotic, because it lacks a nuclear membrane and has a single circular chromosome. The cell is not a bacterium because it has histone proteins, which are present in archaea and eukaryotes, but lacking in bacteria.

Section 2.2

19. A certain species has three pairs of chromosomes: an acrocentric pair, a metacentric pair, and a submetacentric pair. Draw a cell of this species as it would appear in metaphase of mitosis.

Solution:

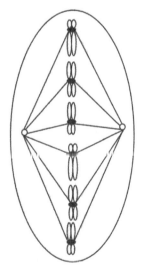

*20. A biologist examines a series of cells and counts 160 cells in interphase, 20 cells in prophase, 6 cells in prometaphase, 2 cells in metaphase, 7 cells in anaphase, and 5 cells in telophase. If the complete cell cycle requires 24 hours, what is the average duration of M phase in these cells? Of metaphase?

Solution:
To determine the average duration of M phase in these cells, the proportion of cells in interphase, or in each stage of M phase, should be calculated by dividing the number of cells in each stage by the total number of cells counted. To calculate the time required for

a given phase, multiply 24 hours by the proportion of cells at that stage. This will give the average duration of each stage in hours.

Stage	Number of cells counted	Proportion of cells at each stage	Average duration (hours)
Interphase	160	0.80	19.2
Prophase	20	0.10	2.4
Prometaphase	6	0.03	0.72
Metaphase	2	0.01	0.24
Anaphase	7	0.035	0.84
Telophase	5	0.025	0.6
Totals	200	1.0	24

The average duration of M phase can be determined by adding up the hours spent in each stage of mitosis. In these cells, M phase lasts 4.8 hours. The table shows that metaphase requires 0.24 hours, or 14.4 minutes.

Section 2.3

21.   A certain species has three pairs of chromosomes: one acrocentric pair and two metacentric pairs. Draw a cell of this species as it would appear in the following stages of meiosis.

a. Metaphase I

Solution:

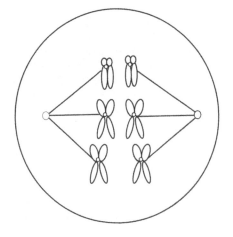

b. Anaphase I

Solution:

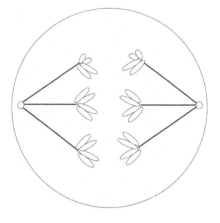

c. Metaphase II

Solution:

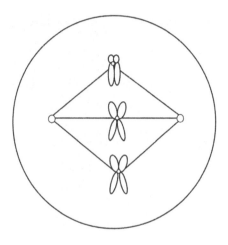

d. Anaphase II

Solution:

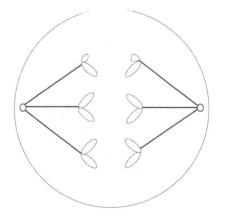

22.   Construct a table similar to that in **Figure 2.10** for the different stages of meiosis, giving the number of chromosomes per cell and the number of DNA molecules per cell for a cell that begins with four chromosomes (two homologous pairs) in G_1. Include the following stages in your table: G_1, S, G_2, prophase I, metaphase I, anaphase I, telophase I (after cytokinesis), prophase II, metaphase II, anaphase II, and telophase II (after cytokinesis).

Solution:

	G1	S	G2	P	MI	AI	TI	PII	MII	AII	TII
Number of chromosomes per cell	4	4	4	4	4	4	2	2	2	4	2
Number of DNA molecules per cell	4	4 to 8	8	8	8	8	4	4	4	4	2

*23.   A cell in G_1 of interphase has 12 chromosomes. How many chromosomes and DNA molecules will be found per cell when this original cell progresses to the following stages?

Solution:
The number of chromosomes and DNA molecules depends on the stage of the cell cycle. Each chromosome contains only one centromere, but after the completion of S phase, and prior to anaphase of mitosis or anaphase II of meiosis, each chromosome will consist of two DNA molecules.

a. G_2 of interphase

Solution:
12 chromosomes and 24 DNA molecules

b.   Metaphase I of meiosis

Solution:
12 chromosomes and 24 DNA molecules

c.   Prophase of mitosis

Solution:
12 chromosomes and 24 DNA molecules

d.   Anaphase I of meiosis

Solution:
12 chromosomes and 24 DNA molecules

e.   Anaphase II of meiosis

Solution:
12 chromosomes and 12 DNA molecules

f. Prophase II of meiosis

Solution:
6 chromosomes and 12 DNA molecules

g. After cytokinesis following mitosis

Solution:
12 chromosomes and 12 DNA molecules

h. After cytokinesis following meiosis II

Solution:
6 chromosomes and 6 DNA molecules

24.   How are the events that take place in spermatogenesis and oogenesis similar? How are they different?

Solution:
Both spermatogenesis and oogenesis begin similarly in that the diploid primordial cells (spermatogonia and oogonia) can undergo multiple rounds of mitosis to produce more primordial cells, or both types of cells can enter into meiotic division. In spermatogenesis, cytokinesis is equal, resulting in haploid cells of similar sizes. Upon completion of meiosis II, four haploid spermatids have been produced for each spermatogonium that began meiosis. In oogenesis, cytokinesis is unequal. At the completion of meiosis I in oogenesis, a secondary oocyte is produced, which is much larger and contains more cytoplasm than the other haploid cell produced, called the first polar body. At the completion of meiosis II, the secondary oocyte divides, producing the ovum and the second polar body. Again, the division of the cytoplasm in cytokinesis is unequal, with the ovum receiving most of the cytoplasmic material. Usually, the polar bodies disintegrate, leaving the ovum as the only product of meiosis.

*25.   All of the following cells, shown in various stages of mitosis and meiosis, come from the same rare species of plant.

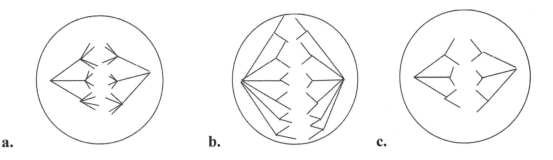

a.   **b.**   **c.**

a. What is the diploid number of chromosomes in this plant?
b. Give the names of each stage of mitosis or meiosis shown.

c. Give the number of chromosomes and number of DNA molecules per cell present at each stage.

Solution:

a. 6 chromosomes

b. First cell: anaphase I of meiosis
    Second cell: anaphase of mitosis
    Third cell: anaphase II of meiosis

c. First cell: 6 chromosomes and 12 DNA molecules
    Second cell: 12 chromosomes and 12 DNA molecules
    Third cell: 6 chromosomes and 6 DNA molecules

*26. The amount of DNA per cell of a particular species is measured in cells found at various stages of meiosis, and the following amounts are obtained:

Amount of DNA per cell in picograms (pg)

_____ 3.7 pg _____ 7.3 pg _____ 14.6 pg

Match the amounts of DNA above with the corresponding stages of meiosis (*a* through *f*, below). You may use more than one stage for each amount of DNA.

Stage of meiosis
a. G_1
b. Prophase I
c. G_2
d. Following telophase II and cytokinesis
e. Anaphase I
f. Metaphase II

Solution:

___d___ 3.7 pg __a, f__ 7.3 pg __b, c, e__ 14.6 pg

The amount of DNA in the cell will be doubled after the completion of S phase in the cell cycle and prior to cytokinesis in either mitosis or meiosis I. At the completion of cytokinesis following meiosis II, the amount of DNA will be halved.

a. G_1 occurs prior to S phase and the doubling of the amount of DNA and prior to the completion of the meiosis II and cytokinesis, which will result in a haploid cell containing one-half the amount of DNA that was contained in the cell in G_1.
b. During prophase I of meiosis, the amount of DNA in the cell is two times the amount in G_1. The homologous chromosomes are still located within a single cell, and there are two sister chromatids per chromosome.

c. G_2 takes place directly after the completion of the S phase, so the amount of DNA is two times the amount prior to the S phase.

d. Following cytokinesis associated with meiosis II, each daughter cell will contain only one-half the amount of DNA of a mother cell found in G_1 of interphase. By the completion of cytokinesis associated with meiosis II, both homologous pairs of chromosomes and sister chromatids have been separated into different daughter cells. Therefore, each daughter cell will contain only one-half the amount of DNA of the original cell in G_1.

e. During anaphase I of meiosis, the amount of DNA in the cell is two times the amount in G_1. The homologous chromosomes are still located within a single cell, and there are two sister chromatids per chromosome.

f. Metaphase II takes place after the cytokinesis associated with meiosis I and results in the daughter cells receiving only one-half the DNA found in their mother cell. In metaphase II of meiosis, the amount of DNA in each cell is the same as G_1 because each chromosome still consists of two DNA molecules (two sister chromatids per chromosome).

27. A cell in prophase II of meiosis has 12 chromosomes. How many chromosomes would be present in a cell from the same organism if it were in prophase of mitosis? Prophase I of meiosis?

   Solution:
   A cell in prophase II of meiosis will contain the haploid number of chromosomes. For this organism, 12 chromosomes represent the haploid chromosome number of a cell, or one complete set of chromosomes.

   A cell from the same organism that is undergoing prophase of mitosis would contain a diploid number of chromosomes, or two complete sets of chromosomes, which means that homologous pairs of chromosomes are present. So a cell in this stage should contain 24 chromosomes.

   Homologous pairs of chromosomes have not been separated by prophase I of meiosis. During this stage, a cell of this organism will contain 24 chromosomes.

28. A cell has 8 chromosomes in G_1 of interphase. Draw a picture of this cell with its chromosomes at the following stages. Indicate how many DNA molecules are present at each stage.

Solution:

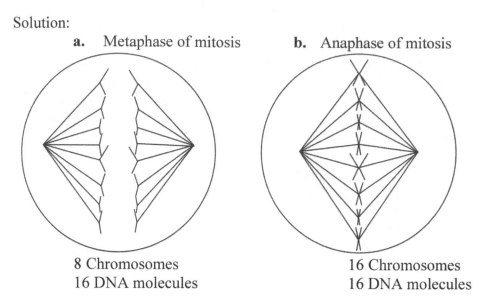

a.   Metaphase of mitosis   **b.**   Anaphase of mitosis

8 Chromosomes   16 Chromosomes
16 DNA molecules   16 DNA molecules

c.   Anaphase II of meiosis

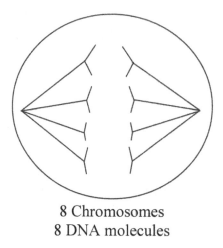

8 Chromosomes
8 DNA molecules

*29.   The fruit fly *Drosophila melanogaster* has four pairs of chromosomes, whereas the house fly *Musca domestica* has six pairs of chromosomes. In which species would you expect to see more genetic variation among the progeny of a cross? Explain your answer.

Solution:
The progeny of an organism whose cells contain the larger number of homologous pairs of chromosomes should be expected to exhibit more variation. The number of different combinations of chromosomes that are possible in the gametes is 2^n, where n is equal to the number of homologous pairs of chromosomes. For the fruit fly, which has four pairs of chromosomes, the number of possible combinations is $2^4 = 16$. For the house fly, which has six pairs of chromosomes, the number of possible combinations is $2^6 = 64$.

*30. A cell has two pairs of submetacentric chromosomes, which we will call chromosomes I_a, I_b, II_a, and II_b (chromosomes I_a and I_b are homologs, and chromosomes II_a and II_b are homologs). Allele M is located on the long arm of chromosome I_a, and allele m is located at the same position on chromosome I_b. Allele P is located on the short arm of chromosome I_a, and allele p is located at the same position on chromosome I_b. Allele R is located on chromosome II_a, and allele r is located at the same position on chromosome II_b.

    a. Draw these chromosomes, identifying genes M, m, P, p, R, and r as they might appear in metaphase I of meiosis. Assume that there is no crossing over.

    Solution:

Metaphase I

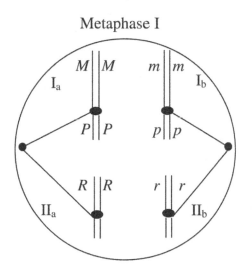

    b. Taking into consideration the random separation of chromosomes in anaphase I, draw the chromosomes (with genes identified) present in all possible types of gametes that might result from this cell's undergoing meiosis. Assume that there is no crossing over.

    Solution:

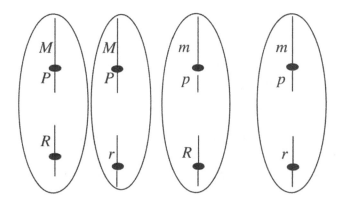

31. A horse has 64 chromosomes and a donkey has 62 chromosomes. A cross between a female horse and a male donkey produces a mule, which is usually sterile. How many chromosomes does a mule have? Can you think of any reasons for the fact that most mules are sterile?

Solution:
The haploid egg produced by the female horse contains 32 chromosomes. The haploid sperm produced by the male donkey contains 31 chromosomes. The union of the horse and donkey gametes will produce a zygote containing 63 chromosomes. From the zygote, the adult mule will develop and will contain cells with a chromosome number of 63. Because an odd number of chromosomes in the mule's cells are present, at least one chromosome will not have a homolog. During the production of gametes by meiosis when pairing and separation of homologous chromosomes occurs, the odd chromosome will be unable to pair up. Furthermore, the mule's chromosomes, which are contributed by the horse and donkey, are from two different species. Not all of the mule's chromosomes may be able to find a suitable homolog during meiosis I and thus may not synapse properly during prophase I of meiosis. If improper synapsis or no synapsis occurs during prophase I, this will result in faulty segregation of chromosomes to the daughter cells produced at the conclusion of meiosis I. This leads to gametes that have abnormal numbers of chromosomes. When these abnormal gametes unite, the resulting zygote has an abnormal number of chromosomes and will be nonviable.

*32. Normal somatic cells of horses have 64 chromosomes ($2n = 64$). How many chromosomes and DNA molecules will be present in the following types of horse cells?

Solution:

Cell type	Number of chromosomes	Number of DNA molecules
a. Spermatogonium	64	64

Assuming the spermatogonium is in G_1 prior to the production of sister chromatids in S phase, the chromosome number will be the diploid number of chromosomes.

b. First polar body	32	64

The first polar body is the product of meiosis I, so it will be haploid; but the sister chromatids have not separated, so each chromosome will consist of two sister chromatids.

c. Primary oocyte	64	128

The primary oocyte has stopped in prophase I of meiosis. So the homologs have not yet separated, and each chromosome consists of two sister chromatids.

d. Secondary spermatocyte	32	64

The secondary spermatocyte is a product of meiosis I and has yet to enter meiosis II. So the secondary spermatocyte will be haploid because the homologous pairs were separated in meiosis I; but each chromosome is still composed of two sister chromatids.

33.   Indicate whether each of the following cells is haploid or diploid.

Solution:

Cell type	Haploid or diploid?
Primary spermatocyte	diploid
Microsporocyte	diploid
First polar body	haploid
Oogonium	diploid
Spermatid	haploid
Megaspore	haploid
Ovum	haploid
Secondary oocyte	haploid
Spermatogonium	diploid

*34.   A primary oocyte divides to give rise to a secondary oocyte and a first polar body. The secondary oocyte then divides to give rise to an ovum and a second polar body.

a.   Is the genetic information found in the first polar body identical with that found in the secondary oocyte? Explain your answer.

Solution:
No, the information is not identical with that found in the secondary oocyte. The first polar body and the secondary oocyte are the result of meiosis I. In meiosis I, homologous chromosomes segregate and thus both the first polar body and secondary oocyte will contain only one member of each original chromosome pair, and these will have different alleles of some of the genes. Also, the recombination that took place in prophase I will have generated new and different arrangements of genetic material for each member of the pair.

b.   Is the genetic information found in the second polar body identical with that in the ovum? Explain your reasoning.

Solution:
No, the information is not identical. The second polar body and the ovum will contain the same members of the homologous pairs of chromosomes that were separated during meiosis I and produced by the separation of sister chromatids during anaphase II. However, the sister chromatids are no longer identical. The sister chromatids have undergone recombination during prophase I and thus contain genetic information that is not identical to the other sister chromatids.

CHALLENGE QUESTIONS

Section 2.3

*35. Female bees are diploid, and male bees are haploid. The haploid males produce sperm and can successfully mate with diploid females. Fertilized eggs develop into females and unfertilized eggs develop into males. How do you think the process of sperm production in male bees differs from sperm production in other animals?

Solution:
Most male animals produce sperm by meiosis. Because meiosis takes place only in diploid cells, haploid male bees do not undergo meiosis. Male bees can produce sperm, but only through mitosis. Haploid cells that divide mitotically produce more haploid cells.

36.   On average, what proportion of the genome in the following pairs of humans would be exactly the same if no crossing over occurred? (For the purposes of this question only, we will ignore the special case of the X and Y sex chromosomes and assume that all genes are located on nonsex chromosomes.)

  a. Father and child

  Solution:
  The father will donate one-half of his chromosomes to his child. Therefore, the father and child will have one-half of their genomes that are similar.

  b. Mother and child

  Solution:
  The mother will donate one-half of her chromosomes to her child. Therefore, the mother and child will have one-half of their genomes that are similar.

  c. Two full siblings (offspring that have the same two biological parents)

  Solution:
  The parents can contribute only one-half of their genome to each offspring. So it is likely that the siblings share one-fourth of their genes from one parent. Because each sibling would share one-fourth of their genes from each parent, their total relatedness is one-half (or ¼ + ¼).

  d. Half siblings (offspring that have only one biological parent in common)

  Solution:
  Half siblings share only one-fourth of their genomes with each other because they have only one parent in common.

e. Uncle and niece

Solution:
An uncle would share one-half of his genomes with his sibling, who would share one-half of his or her genome with his or her child. So, an uncle and niece would share one-fourth of their genomes ($\frac{1}{2} \times \frac{1}{2}$).

f. Grandparent and grandchild

Solution:
The grandparent and grandchild would share one-fourth of their genomes because the grandchild would share one-half of her genome with her parent and the parent would share one-half of her genome with the child's grandparent.

Chapter Three: Basic Principles of Heredity

COMPREHENSION QUESTIONS

Section 3.1

1. Why was Mendel's approach to the study of heredity so successful?

    Solution:
    Mendel was successful for several reasons. He chose to work with a plant, *Pisum sativum*, which was easy to cultivate, grew relatively rapidly, and produced many offspring whose phenotype was easy to determine, which allowed Mendel to detect mathematical ratios of progeny phenotypes. The seven characteristics that he chose to study exhibited only a few distinct phenotypes and did not show a range of variation. Finally, by looking at each trait separately and counting the numbers of the different phenotypes, Mendel adopted a reductionist experimental approach and applied the scientific method. From his observations, he proposed hypotheses that he was then able to test empirically.

2. What is the difference between genotype and phenotype?

    Solution:
    Genotype refers to the genes or the set of alleles found within an individual. Phenotype refers to the manifestation of a particular character or trait.

Section 3.2

3. What is the principle of segregation? Why is it important?

    Solution:
    The principle of segregation states that an organism possesses two alleles for any particular characteristic. These alleles separate during the formation of gametes. In other words, one allele goes into each gamete. The principle of segregation is important because it explains how the genotypic ratios in the haploid gametes are produced.

4. How are Mendel's principles different from the concept of blending inheritance discussed in Chapter 1?

    Solution:
    Mendel's principles assert that the genetic factors or alleles are discrete units that remain separate in an individual organism with a trait encoded by the dominant allele being the only one observed if two different alleles are present. According to Mendel's principles, if an individual contains two different alleles, then the individual's gametes could contain either of these two alleles (but not both). Blending inheritance proposes that offspring are the result of blended genetic material from the parent and the genetic factors are not discrete units. Once blended, the combined genetic material could not be separated from each other in future generations.

5.   What is the concept of dominance?

Solution:
The concept of dominance states that when two different alleles are present in a genotype, only the dominant allele is expressed in the phenotype. Incomplete dominance occurs when different alleles are expressed in a heterozygous individual, and the resulting phenotype is intermediate to the phenotypes of the two homozygotes.

6.   What are the addition and multiplication rules of probability and when should they be used?

Solution:
The addition and multiplication rules are two rules of probability used by geneticists to predict the ratios of offspring in genetic crosses. The multiplication rule allows for predicting the probability of two or more independent events occurring together. According to the multiplication rule, the probability of two independent events occurring together is the product of their probabilities of occurring independently. The addition rule allows for predicting the likelihood of a single event that can happen in two or more ways. It states that the probability of a single mutually exclusive event can be determined by adding the probabilities of the two or more different ways in which this single event could take place. The multiplication rule allows us to predict how alleles from each parent can combine to produce offspring, whereas the addition rule is useful in predicting phenotypic ratios once the probability of each type of progeny can be determined.

7.   Give the genotypic ratios that may appear among the progeny of simple crosses and the genotypes of the parents that may give rise to each ratio.

Solution:

Genotypic ratio	Parental genotype
1:2:1	Aa × Aa
1:1	Aa × aa
Uniform progeny	AA × AA aa × aa AA × aa

8.   What is the chromosome theory of heredity? Why was it important?

Solution:
Walter Sutton's chromosome theory of inheritance states that genes are located on the chromosomes. The independent segregation of pairs of homologous chromosomes in meiosis provides the biological basis for Mendel's two principles of heredity.

Section 3.3

9.   What is the principle of independent assortment? How is it related to the principle of segregation?

Solution:
The principle of independent assortment states that alleles at different loci segregate independently of one another. The principle of independent assortment is an extension of the principle of segregation. The principle of segregation states that the two alleles at a locus separate; according to the principle of independent assortment, when these two alleles separate, their separation is independent of the separation of alleles at other loci.

10.   In which phases of mitosis and meiosis are the principles of segregation and independent assortment at work?

Solution:
In anaphase I of meiosis, each pair of homologous chromosomes segregates independently of all other pairs of homologous chromosomes. The assortment is dependent on how the homologs line up during metaphase I. This assortment of homologs explains how genes located on different pairs of chromosomes will separate independently of one another. Anaphase II results in the separation of sister chromatids and subsequent production of gametes carrying single alleles for each gene locus as predicted by Mendel's principle of segregation.

Section 3.4

11.   How is the chi-square goodness-of-fit test used to analyze genetic crosses? What does the probability associated with a chi-square value indicate about the results of a cross?

Solution:
The chi-square goodness-of-fit test is a statistical method used to evaluate the role of chance in causing deviations between the observed and the expected numbers of offspring produced in a genetic cross. The probability value obtained from the chi-square table refers to the probability that random chance produced the deviations of the observed numbers from the expected numbers.

Section 3.5

12.   What features are exhibited by a pedigree of a recessive trait? What features are exhibited if the trait is dominant?

Solution:
Assuming autosomal inheritance and complete penetrance, recessive inheritance is indicated by the presence of affected offspring from two unaffected parents. If the trait is rare, frequently the recessive allele will be passed for a number of generations without the trait appearing in the pedigree. For rare recessive traits, the trait often appears as a result of mating between two close relatives. Finally, if the trait is recessive, all of the offspring of two affected parents will be affected.

Assuming autosomal inheritance and complete penetrance, at least one of the parents of affected children should be affected (unless a new mutation has led to creation of a new dominant allele). The trait should not skip generations within a lineage. Two affected parents who are heterozygotes can have unaffected offspring. Unaffected parents do not transmit the trait to their offspring.

APPLICATION QUESTIONS AND PROBLEMS

Introduction

13. If blond hair in the Solomon Islanders had originated from early European explorers, what would you predict the researchers would have found when they conducted their genetic study of the islanders?

    Solution:
    The genome-wide association studies of the Solomon islanders would have found associations between blond hair and genes that have been shown to be associated with blond hair in northern Europeans. When they sequenced the DNA of blond and dark-haired Solomon islanders they would have found differences in those genes which cause blond hair in Europeans, and the specific mutation that caused blond hair in Solomon islanders would have also been found in Europeans.

Section 3.1

*14. What characteristics of an organism would make it suitable for studies of the principles of inheritance? Can you name several organisms that have these characteristics?

    Solution:
    Useful characteristics
    - Are easy to grow and maintain
    - Grow rapidly, producing many generations in a short period
    - Produce large numbers of offspring
    - Have distinctive phenotypes that are easy to recognize

    Examples of organisms that meet these criteria
    - *Neurospora*, a fungus
    - *Saccharomyces cerevisiae*, a yeast
    - *Arabidopsis*, a plant
    - *Caenorhabditis elegans*, a nematode
    - *Drosophila melanogaster*, a fruit fly

Section 3.2

15. In cucumbers, orange fruit color (R) is dominant over cream fruit color (r). A cucumber plant homozygous for orange fruit is crossed with a plant homozygous for cream fruit. The F_1 are intercrossed (F_1 individuals crossed with F_1 individuals) to produce the F_2.

a. Give the genotypes and phenotypes of the parents, the F₁, and the F₂.

Solution:

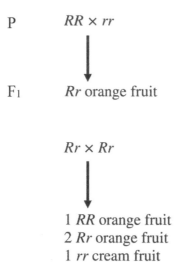

P   *RR* × *rr*

F₁   *Rr* orange fruit

Rr × *Rr*

1 *RR* orange fruit
2 *Rr* orange fruit
1 *rr* cream fruit

The parents are *RR* (orange fruit) and *rr* (cream fruit). All of the F₁ are *Rrm* (orange). The F₂ are 1 *RR*:2 *Rr*:1 *rr* and have an orange-to-cream phenotypic ratio of 3:1.

b. Give the genotypes and phenotypes of the offspring of a backcross between the F₁ and the orange-fruit parent.

Solution:
Half of the progeny are homozygous for orange fruit (*RR*) and half are heterozygous for orange fruit (*Rr*).

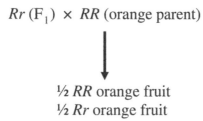

Rr (F₁) × *RR* (orange parent)

½ *RR* orange fruit
½ *Rr* orange fruit

c. Give the genotypes and phenotypes of a backcross between the F₁ and the cream-fruit parent.

Solution:
Half of the progeny are heterozygous for orange fruit (*Rr*) and half are homozygous for cream fruit (*rr*).

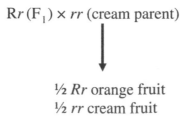

R*r* (F₁) × *rr* (cream parent)

½ *Rr* orange fruit
½ *rr* cream fruit

16.   **Figure 1.1** (p. 2) shows three girls, one of whom has albinism. Could the three girls shown in the photograph be sisters? Why or why not?

Solution:
Yes, the three girls could be sisters. Albinism is an autosomal recessive trait. The girl with albinism phenotype must carry two copies of the albinism allele in order to express the trait. The other two girls could be homozygous for normal pigmentation allele or heterozygous for the albinism allele. In each case, the girls could have the same parent. Both parents could be heterozygous for the albinism allele or one parent could be heterozygous whereas the other parent would be homozygous for the albinism allele. In either circumstance, the parents could have children who have normal pigmentation or exhibit albinism.

17.   [Data Analysis Problem] J. W. McKay crossed a stock melon plant that produced tan seeds with a plant that produced red seeds and obtained the following results (J. W. McKay. 1936. *Journal of Heredity* 27:110–112).

Cross	**F$_1$**	**F$_2$**
tan ♀ × red ♂	13 tan seeds	93 tan, 24 red seeds

a.   Explain the inheritance of tan seeds and red seeds in this plant.

Solution:
The F$_1$ generation contains all tan-seed-producing progeny and is the result of crossing a tan-seed-producing plant with a red-seed-producing plant. The F$_1$ result suggests that the tan phenotype is dominant to red. In the F$_2$ generation, the ratio of tan- to red-seed-producing plants is about 3.9 to 1, which is similar but not identical to a 3 to 1 ratio expected for a monohybrid cross involving dominant and recessive alleles. The F$_2$ ratio suggests that the F$_1$ parents are heterozygous dominant for tan color.

b.   Assign symbols for the alleles in this cross and give genotypes for all the individual plants.

Solution:
We will define the tan allele as "*R*" and the recessive red allele as "*r*."
Tan-seed-producing ♀ parent: *RR*
Red-seed-producing ♂ parent: *rr*
F$_1$ tan-seed-producing offspring: *Rr*
F$_2$ tan-seed-producing offspring: *RR* or *Rr*
F$_2$ red-seed-producing offspring: *rr*

*18.   [Data Analysis Problem] White (*w*) coat color in guinea pigs is recessive to black (*W*). In 1909, W. E. Castle and J. C. Phillips transplanted an ovary from a black guinea pig into a white female whose ovaries had been removed. They then mated this white female with a white male. All the offspring from the mating were black in color (W. E. Castle and J. C. Phillips. 1909. *Science* 30:312–313).

a. Explain results of this cross.

Solution:
Although the white female gave birth to the offspring, her eggs were produced by the ovary from the black female guinea pig. The transplanted ovary produced only eggs containing the allele for black coat color. Like most mammals, guinea pig females produce primary oocytes early in development, and thus the transplanted ovary already contained primary oocytes produced by the black female guinea pig.

b. Give the genotype of the offspring of this cross.

Solution:
The white male guinea pig contributed a "*w*" allele, whereas the white female guinea pig contributed the "*W*" allele from the transplanted ovary. The offspring are thus *Ww*.

c. What, if anything, does this experiment indicate about the validity of the pangenesis and the germ-plasm theories discussed in Chapter 1?

Solution:
The transplant experiment supports the germ-plasm theory. According to the germ-plasm theory, only the genetic information in the germ-line tissue in the reproductive organs is passed to the offspring. The production of black guinea pig offspring suggests that the allele for black coat color was passed to the offspring from the transplanted ovary in agreement with the germ-plasm theory. According to the pangenesis theory, the genetic information passed to the offspring originates at various parts of the body and travels to the reproductive organs for transfer to the gametes. If pangenesis were correct, then the guinea pig offspring should have been white. The white coat alleles would have traveled to the transplanted ovary and then to the white female's gametes. The absence of any white offspring indicates that pangenesis did not occur.

*19.   In cats, blood type A results from an allele I^A which is dominant over an allele i^B that produces blood type B. (There is no O blood type, as there is in humans.) The blood types of male and female cats that were mated and the blood types of their kittens are presented in the following table. Give the most likely genotypes for the parents of each litter.

	Male parent	Female parent	Kittens
a.	A	B	4 with type A, 3 kittens with type B
b.	B	B	6 kittens with type B
c.	B	A	8 kittens with type A
d.	A	A	7 kittens with type A, 2 kittens with type B
e.	A	A	10 kittens with type A
f.	A	B	4 kittens with type A, 1 kitten with type B

a. Male with blood type A × female with blood type B

Solution:
Because the female parent has blood type B, she must have the genotype $i^B i^B$. The male parent could be either $I^A I^A$ or $I^A i^B$. However, as some of the offspring are kittens with blood type B, the male parent must have contributed an i^B allele to these kittens. Therefore, the male must have the genotype of $I^A i^B$.

b. Male with blood type B × female with blood type B

Solution:
Because blood type B is caused by the recessive allele i^B, both parents must be homozygous for the recessive allele or $i^B i^B$. Each contributes only the i^B allele to the offspring.

c. Male with blood type B × female with blood type A

Solution:
Again, the male with type B blood must be $i^B i^B$. A female with type A blood could have either the $I^A I^A$ or $I^A i^B$ genotypes. Because all of her kittens have type A blood, this suggests that she is homozygous for the for I^A allele ($I^A I^A$) and contributes only the I^A allele to her offspring. It is possible that she is heterozygous for type A blood, but if so it is unlikely that chance alone would have produced eight kittens with blood type A.

d. Male with blood type A × female with blood type A

Solution:
Because kittens with blood type A and blood type B are found in the offspring, both parents must be heterozygous for blood type A, or $I^A i^B$. With both parents being heterozygous, the offspring would be expected to occur in a 3:1 ratio of blood type A to blood type B, which is close to the observed ratio.

e. Male with blood type A × female with blood type A

Solution:
Only kittens with blood type A are produced, which suggests that each parent is homozygous for blood type A ($I^A I^A$), or that one parent is homozygous for blood type A ($I^A I^A$) and the other parent is heterozygous for blood type A ($I^A i^B$). The data from the offspring will not allow us to determine the precise genotype of either parent.

f. Male with blood type A × female with blood type B

Solution:
On the basis of her phenotype, the female will be $i^B i^B$. In the offspring, one kitten with blood type B is produced. This kitten would require that both parents contribute an i^B to produce its genotype. Therefore, the male parent's genotype is $I^A i^B$. From this cross,

the number of kittens with blood type B would be expected to be similar to the number of kittens with blood type A. However, due to the small number of offspring produced, random chance could have resulted in more kittens with blood type A than kittens with blood type B.

20.   **Figure 3.7** shows the results of a cross between a tall pea plant and a short pea plant.

   a.   What phenotypes will be produced, and in what proportions, if a tall F₁ plant is backcrossed to the short parent?

   Solution:
   The tall F₁ plants are all heterozygous (*Tt*). The short parent is homozygous for the short allele (*tt*).

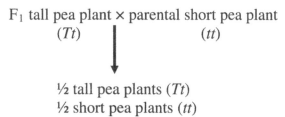

$$F_1 \text{ tall pea plant} \times \text{parental short pea plant}$$
$$(Tt) \qquad\qquad\qquad (tt)$$

½ tall pea plants (*Tt*)
½ short pea plants (*tt*)

   b.   What phenotypes will be produced, and in what proportions, if a tall F₁ plant is backcrossed to the tall parent?

   Solution:
   The tall F₁ plants are all heterozygous (*Tt*). The tall parent is heterozygous (*Tt*) as well.

$$F_1 \text{ tall pea plant} \times \text{parental tall pea plant}$$
$$(Tt) \qquad\qquad\qquad (Tt)$$

¾ tall pea plants (*TT* and *Tt*)
¼ short pea plants (*tt*)

21.   Joe has a white cat named Sam. When Joe crosses Sam with a black cat, he obtains ½ white kittens and ½ black kittens. When the black kittens are interbred, all the kittens that they produce are black. On the basis of these results, which coat color (white or black) in cats would you conclude is a recessive trait? Explain your reasoning.

   Solution:
   The black coat color is probably recessive. When Sam was crossed with a black cat, one-half the offspring were white and one-half were black. This ratio potentially indicates that one of the parental cats is heterozygous dominant whereas the other parental cat is homozygous recessive—a testcross. The interbreeding of the black kittens produced only black kittens, indicating that the black kittens are probably to be homozygous, and thus the black coat color is the recessive trait.

If the black allele was dominant, we would have expected the black kittens to be heterozygous, containing a black coat color allele and a white coat color allele. Under this condition, we would expect one-fourth of the progeny from the interbred black kittens to have white coats. Because this did not happen, we can conclude that the black coat color is recessive.

*22.   Alkaptonuria is a metabolic disorder in which affected persons produce black urine. Alkaptonuria results from an allele (*a*) that is recessive to the allele for normal metabolism (*A*). Sally has normal metabolism, but her brother has alkaptonuria. Sally's father has alkaptonuria, and her mother has normal metabolism.

    a.   Give the genotypes of Sally, her mother, her father, and her brother.

        Solution:
        Sally's father, who has alkaptonuria, must be *aa*. Her brother, who also has alkaptonuria, must be *aa* as well. Because both parents must have contributed one *a* allele to her brother, Sally's mother, who is phenotypically normal, must be heterozygous (*Aa*). Sally, who is normal, received the *A* allele from her mother but must have received an *a* allele from her father.
        The genotypes of the individuals are: Sally (*Aa*), Sally's mother (*Aa*), Sally's father (*aa*), and Sally's brother (*aa*).

    b.   If Sally's parents have another child, what is the probability that this child will have alkaptonuria?

        Solution:
        Sally's father (*aa*) × Sally's mother (*Aa*)
        Sally's mother has a one-half chance of contributing the *a* allele to her offspring. Sally's father can contribute only the *a* allele. The probability of an offspring with genotype *aa* and alkaptonuria is therefore ½ × 1 = ½.

    c.   If Sally marries a man with alkaptonuria, what is the probability that their first child will have alkaptonuria?

        Solution:
        Since Sally is heterozygous (*Aa*), she has a one-half chance of contributing the *a* allele. Her husband with alkaptonuria (*aa*) can only contribute the *a* allele. The probability of their first child (as well as for any additional child) having alkaptonuria (*aa*) is ½ × 1 = ½.

23.   Hairlessness in American rat terriers is recessive to the presence of hair. Suppose that you have a rat terrier with hair. How can you determine whether this dog is homozygous or heterozygous for the hairy trait?

Solution:
Use *h* for the recessive hairless allele and *H* for the dominant allele for the presence of hair. Because *H* is dominant over *h*, a rat terrier with hair could be either homozygous (*HH*) or heterozygous (*Hh*). To determine which genotype is present in the rat terrier with hair, cross this dog with a hairless rat terrier (*hh*). If the terrier with hair is homozygous (*HH*), then no hairless offspring will be produced by the testcross. However, if the terrier is heterozygous (*Hh*), then 1/2 of the offspring will be hairless.

*24.   What is the probability of rolling one six-sided die and obtaining the following numbers?

   a. 2

   Solution:
   Because 2 is only found on one side of a six-sided die, then there is a 1/6 chance of rolling a two.

   b. 1 or 2

   Solution:
   The probability of rolling a 1 on a six-sided die is 1/6. Similarly, the probability of rolling a 2 on a six-sided die is 1/6. Because the question asks what is the probability of rolling a 1 or a 2, and these are mutually exclusive events, we should use the additive rule of probability to determine the probability of rolling a 1 or a 2: (probability of rolling a 1) + (probability of rolling a 2) = probability of rolling either a 1 or a 2
   1/6 + 1/6 = 2/6 = 1/3 probability of rolling either a 1 or a 2

   c. An even number

   Solution:
   The probability of rolling an even number depends on the number of even numbers found on the die. A single die contains three even numbers (2, 4, 6). The probability of rolling any one of these three numbers on a six-sided die is 1/6. To determine the probability of rolling either a 2, a 4, or a 6, we apply the additive rule: 1/6 + 1/6 + 1/6 = 3/6 = ½.

   d. Any number but a 6

   Solution:
   The number 6 is found only on one side of a six-sided die. The probability of rolling a 6 is therefore 1/6. The probability of rolling any number but 6 is (1 – 1/6) = 5/6.

*25. What is the probability of rolling two six-sided dice and obtaining the following numbers?

    a. 2 and 3

    Solution:
    1/18

    b. 6 and 6

    Solution:
    1/36

    c. At least one 6

    Solution:
    11/36

    d. Two of the same number (two 1s, or two 2s, or two 3s, etc.)

    Solution:
    1/6

    e. An even number on both dice

    Solution:
    ¼

    f. An even number on at least one die

    Solution:
    ¾

26. Phenylketonuria (PKU) is a disease that results from a recessive gene. Two normal parents produce a child with PKU.

    a. What is the probability that a sperm from the father will contain the PKU allele?

    Solution:
    The father has a ½ chance of donating a sperm with the PKU allele.

    b. What is the probability that an egg from the mother will contain the PKU allele?

    Solution:
    The mother's egg has a ½ chance of containing the PKU allele.

c. What is the probability that their next child will have PKU?

Solution:
Each parent has a ½ chance of donating the *p* allele to the child. So, the child has a ½ × ½ = ¼ chance of having PKU.

d. What is the probability that their next child will be heterozygous for the PKU gene?

Solution:
Each parent has a ½ chance of donating the *P* allele or a ½ chance of donating the *p* allele to the child. Therefore, the child has a (½ × ½) + (½ × ½) = ½ chance of being heterozygous.

*27. In German cockroaches, curved wing (*cv*) is recessive to normal wing (*cv*⁺). A homozygous cockroach that has normal wings is crossed with a homozygous cockroach that has curved wings. The F₁ are intercrossed to produce the F₂. Assume that the pair of chromosomes containing the locus for wing shape is metacentric. Draw this pair of chromosomes as it would appear in the parents, the F₁, and each class of F₂ progeny at metaphase I of meiosis. Assume that no crossing over takes place. At each stage, label a location for the alleles for wing shape (*cv* and *cv*⁺) on the chromosomes.

Solution:

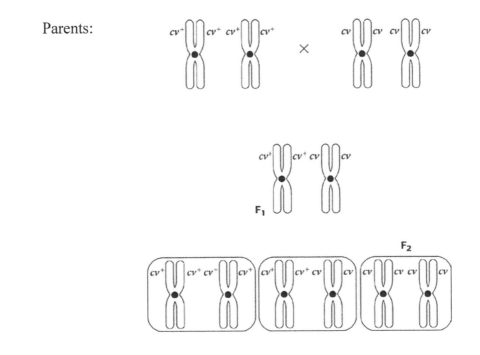

*28. In guinea pigs, the allele for black fur (*B*) is dominant over the allele for brown fur (*b*). A black guinea pig is crossed with a brown guinea pig, producing five F₁ black guinea pigs and six F₁ brown guinea pigs.

a. How many copies of the black allele (*B*) will be present in each cell of an F_1 black guinea pig at the following stages: G_1, G_2, metaphase of mitosis, metaphase I of meiosis, metaphase II of meiosis, and after the second cytokinesis following meiosis? Assume that no crossing over takes place.

Solution:
In the F_1 black guinea pigs (*Bb*), only one chromosome possesses the black allele, so the number of copies present at each stage are as follows: G_1, one black allele; G_2, two black alleles; metaphase of mitosis, two black alleles; metaphase I of meiosis, two black alleles; metaphase II of meiosis, two but only in one-half of the cells—the remaining one-half will not contain the B allele; after cytokinesis of meiosis, one black allele but only in half of the cells produced by meiosis. (The remaining half will not contain the black allele.)

b. How many copies of the brown allele (*b*) will be present in each cell from an F_1 brown guinea pig at the same stages as those listed in part *a*? Assume that no crossing over takes place.

Solution:
In the F_1 brown guinea pigs (*bb*), both homologs possess the brown allele, so the number of copies present at each stage are as follows: G_1, two brown alleles; G_2, four brown alleles; metaphase of mitosis, four brown alleles; metaphase I of meiosis, four brown alleles; metaphase II of meiosis, two brown alleles; after cytokinesis of meiosis, one brown allele.

Section 3.3

29. In watermelons, bitter fruit (*B*) is dominant over sweet fruit (*b*), and yellow spots (*S*) are dominant over no spots (*s*). The genes for these two characteristics assort independently. A homozygous plant that has bitter fruit and yellow spots is crossed with a homozygous plant that has sweet fruit and no spots. The F_1 are intercrossed to produce the F_2.

a. What are the phenotypic ratios in the F_2?

Solution:
P: homozygous bitter fruit and yellow spots (*BB SS*) × homozygous sweet fruit and no spots (*bb ss*)
F_1: All progeny have bitter fruit and yellow spots (*Bb Ss*).
The F_1 are intercrossed to produce the F_2: *Bb Ss* × *Bb Ss*.
The F_2 phenotypic ratios are as follows:
    9/16 bitter fruit and yellow spots
    3/16 bitter fruit and no spots
    3/16 sweet fruit and yellow spots
    1/16 sweet fruit and no spots

b. If an F₁ plant is backcrossed with the bitter, yellow-spotted parent, what phenotypes and proportions are expected in the offspring?

Solution:
The backcross of an F₁ plant (*Bb Ss*) with the bitter, yellow-spotted parent (*BB SS*) will produce all bitter, yellow-spotted offspring.

c. If an F₁ plant is backcrossed with the sweet, nonspotted parent, what phenotypes and proportions are expected in the offspring?

Solution:
The backcross of a F₁ plant (*Bb Ss*) with the sweet, nonspotted parent (*bb ss*) will produce the following phenotypic proportions in the offspring:
   ¼ bitter fruit and yellow spots
   ¼ bitter fruit and no spots
   ¼ sweet fruit and yellow spots
   ¼ sweet fruit and no spots

30.   **Figure 3.9** shows the results of a dihybrid cross involving seed shape and seed color.

a. What proportion of the round and yellow F₂ progeny from this cross is homozygous at both loci?

Solution:
From the Punnett square in **Figure 3.9c**, we can determine that 1/9 of the round and yellow F₂ progeny would be expected to be homozygous at both loci.

b. What proportion of the round and yellow F₂ progeny from this cross is homozygous at least at one locus?

Solution:
From the Punnett square in **Figure 3.9c**, we can determine that 5/9 of the round and yellow F₂ progeny would be expected to be homozygous at least at one locus.

*31.   In cats, curled ears result from an allele (*Cu*) that is dominant over an allele for normal ears (*cu*). Black color results from an independently assorting allele (*G*) that is dominant over an allele for gray (*g*). A gray cat homozygous for curled ears is mated with a homozygous black cat with normal ears. All the F₁ cats are black and have curled ears.

a. If two of the F₁ cats mate, what phenotypes and proportions are expected in the F₂?

Solution:
The parents are *gg Cu Cu* × *GG cu cu*, which produces F₁ progeny with genotype *Gg Cu cu*. If two F₁ cats mated (*Gg Cu cu* × *Gg Cu cu*), the phenotypic ratios expected in the F₂ are as follows:
⁹/₁₆ black with curly ears

$^3/_{16}$ black with normal ears
$^3/_{16}$ gray with curly ears
$^1/_{16}$ gray with normal ears

b. An F₁ cat mates with a stray cat that is gray and possesses normal ears. What phenotypes and proportions of progeny are expected from this cross?

Solution:
The mating of an F₁ cat (*Gg Cu cu*) with a gray cat with normal ears (*gg cu cu*) is expected to produce the following progeny:
¼ black with curly ears
¼ black with normal ears
¼ gray with curly ears
¼ gray with normal ears

*32. The following two genotypes are crossed: *Aa Bb Cc dd Ee × Aa bb Cc Dd Ee.* What will the proportion of the following genotypes be among the progeny of this cross?

a. *Aa Bb Cc Dd Ee*

Solution:
The simplest procedure for determining the proportion of a particular genotype in the offspring is to break the cross down into simple crosses and consider the proportion of the offspring for each cross.

AaBbCcddEe × AabbCcDdEe
Locus 1: *Aa × Aa* = ¼ *AA*, ½ *Aa*, ¼ *aa*
Locus 2: *Bb × bb* = ½ *Bb*, ½ *bb*
Locus 3: *Cc × Cc* = ¼ *CC*, ½ *Cc*, ¼ *cc*
Locus 4: *dd × Dd* = ½ *Dd*, ½ *dd*
Locus 5: *Ee × Ee* = ¼ *EE*, ½ *Ee*, ¼ *ee*

½ (*Aa*) × ½ (*Bb*) × ½ (*Cc*) × ½ (*Dd*) × ½ (*Ee*) = 1/32

b. *Aa bb Cc dd ee*

Solution:

½ (*Aa*) × ½ (*bb*) × ½ (*Cc*) × ½ (*dd*) × ¼ (*ee*) = 1/64

c. *aa bb cc dd ee*

Solution:

¼ (*aa*) × ½ (*bb*) × ¼ (*cc*) × ½ (*dd*) × ¼ (*ee*) = 1/256

d. *AA BB CC DD EE*

Solution:
This will not occur. The *AaBbCcddEe* parent cannot contribute a *D* allele, and the *AabbCcDdEe* parent cannot contribute a *B* allele. Therefore, their offspring cannot be homozygous for the *BB* and *DD* gene loci.

33.   In cucumbers, dull fruit (*D*) is dominant over glossy fruit (*d*), orange fruit (*R*) is dominant over cream fruit (*r*), and bitter cotyledons (*B*) are dominant over nonbitter cotyledons (*b*). The three characters are encoded by genes located on different pairs of chromosomes. A plant homozygous for dull, orange fruit and bitter cotyledons is crossed with a plant that has glossy, cream fruit and nonbitter cotyledons. The F_1 are intercrossed to produce the F_2.

a. Give the phenotypes and their expected proportions in the F_2.

Solution:
All of the F_1 plants have dull, orange fruit and bitter cotyledons (*DdRrBb*). By intercrossing the F_1, the F_2 are produced. The expected phenotypic ratios in the F_2 can be calculated more easily by examining the phenotypic ratios produced by the individual crosses of each gene locus.

F_1 are intercrossed: *DdRrBb* × *DdRrBb*
Locus 1: *Dd* × *Dd* = ¾ dull (*DD* and *Dd*); ¼ glossy (*dd*)
Locus 2: *Rr* × *Rr* = ¾ orange (*RR* and *Rr*); ¼ cream (*rr*)
Locus 3: *Bb* × *Bb* = ¾ bitter (*BB* and *Bb*); ¼ nonbitter (*bb*)

Dull, orange, bitter: ¾ dull × ¾ orange × ¾ bitter = 27/64
Dull, orange, nonbitter: ¾ dull × ¾ orange × ¼ nonbitter = 9/64
Dull, cream, bitter: ¾ dull × ¼ cream × ¾ bitter = 9/64
Dull, cream, nonbitter: ¾ dull × ¼ cream × ¼ nonbitter = 3/64
Glossy, orange, bitter: ¼ glossy × ¾ orange × ¾ bitter = 9/64
Glossy, orange, nonbitter: ¼ glossy × ¾ orange × ¼ nonbitter = 3/64
Glossy, cream, bitter: ¼ glossy × ¼ cream × ¾ bitter = 3/64
Glossy, cream, nonbitter: ¼ glossy × ¼ cream × ¼ nonbitter = 1/64

b. An F_1 plant is crossed with a plant that has glossy, cream fruit and nonbitter cotyledons. Give the phenotypes and expected proportions among the progeny of this cross.

Solution:
Intercrossing the F_1 with a plant that has glossy, cream fruit and nonbitter cotyledons is an example of a testcross. All progeny classes will be expected in equal proportions because the phenotype of the offspring will be determined by the alleles contributed by the F_1 parent.

DdRrCc (F₁) × *ddrrcc* (tester)
F₁ Locus 1 (*Dd*): ½ *D* and ½ *d*
F₁ Locus 2 (*Rr*): ½ *R* and ½ *r*
F₁ Locus 3 (*Cc*): ½ *C* and ½ *c*

Dull, orange, bitter: ½ dull × ½ orange × ½ bitter = 1/8
Dull, orange, nonbitter: ½ dull × ½ orange × ½ nonbitter = 1/8
Dull, cream, bitter: ½ dull × ½ cream × ½ bitter = 1/8
Dull, cream, nonbitter: ½ dull × ½ cream × ½ nonbitter = 1/8
Glossy, orange, bitter: ½ glossy × ½ orange × ½ bitter = 1/8
Glossy, orange, nonbitter: ½ glossy × ½ orange × ½ nonbitter = 1/8
Glossy, cream, bitter: ½ glossy × ½ cream × ½ bitter = 1/8
Glossy, cream, nonbitter: ½ glossy × ½ cream × ½ nonbitter = 1/8

*34. Alleles *A* and *a* are located on a pair of metacentric chromosomes. Alleles *B* and *b* are located on a pair of acrocentric chromosomes. A cross is made between individuals having the following genotypes: *Aa Bb* × *aa bb*.

a. Draw the chromosomes as they would appear in each type of gamete produced by the individuals of this cross.

Solution:
Gametes from *Aa Bb* individual:

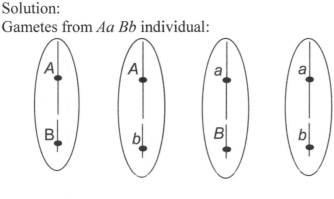

Gametes from *aa bb* individual:

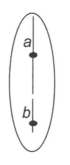

b. For each type of progeny resulting from this cross, draw the chromosomes as they would appear in a cell at G₁, G₂, and metaphase of mitosis.

Solution:
Progeny at G₁:

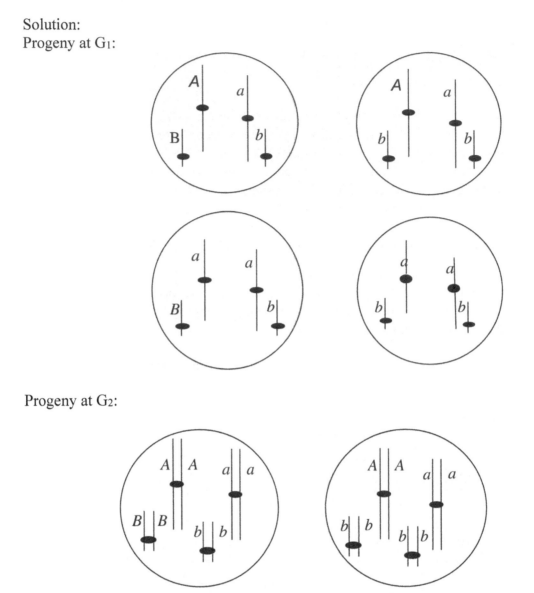

Progeny at G₂:

Progeny at metaphase of mitosis:

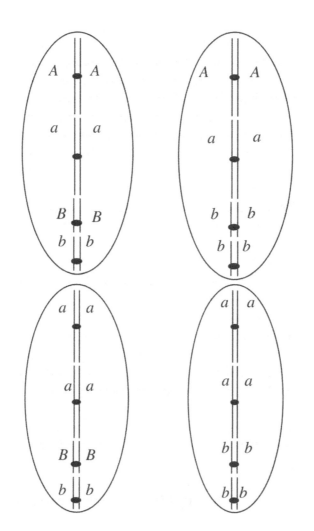

The order of chromosomes on metaphase plate can vary.

Section 3.4

*35. [Data Analysis Problem] J. A. Moore investigated the inheritance of spotting patterns in leopard frogs (J. A. Moore. 1943. *Journal of Heredity* 34:3–7). The pipiens phenotype has the normal spots that give leopard frogs their name. In contrast, the burnsi phenotype lacked spots on its back. Moore carried out the following crosses, producing the progeny indicated.

Parent phenotypes	Progeny phenotypes
burnsi × burnsi	39 burnsi, 6 pipiens
burnsi × pipiens	23 burnsi, 33 pipiens
burnsi × pipiens	196 burnsi, 210 pipiens

a. On the basis of these results, what is the most likely mode of inheritance of the burnsi phenotype?

Solution:
The burnsi × burnsi cross produced both burnsi and pipiens offspring, suggesting that the parents were heterozygous, each possessing a burnsi allele and a pipiens allele. The cross also suggests that the burnsi allele is dominant over the pipiens allele. The progeny of the burnsi × pipiens crosses suggests that each of the crosses was between a homozygous recessive frog (pipiens) and a heterozygous dominant frog (burnsi). The results of both crosses are consistent with the burnsi phenotype being recessive to the pipiens phenotype.

b. Give the most likely genotypes of the parent in each cross (use B for the burnsi allele and B^+ for the pipiens allele).

Solution:
Let B represent the burnsi allele and B^+ represent the pipiens allele.
burnsi ($\underline{BB^+}$) × burnsi ($\underline{BB^+}$)
burnsi ($\underline{BB^+}$) × pipiens (B^+B^+)
burnsi ($\underline{BB^+}$) × pipiens (B^+B^+)

c. Use a chi-square test to evaluate the fit of the observed numbers of progeny to the number expected on the basis of your proposed genotypes.

Solution:
For the burnsi × burnsi cross ($\underline{BB^+} \times BB^+$), we would expect a phenotypic ratio of 3:1 in the offspring.

Phenotype	Observed (O)	Expected (E)	$(O{-}E)^2/E$ or (χ^2)
burnsi	39	33.75	0.81
pipiens	6	11.25	2.45
Total	45	45	3.26

A chi-square test to evaluate the fit of the observed numbers of progeny with an expected 3:1 ratio gives a chi-square value of 2.706 with 1 degree of freedom. The probability associated with this chi-square value is between 0.1 and 0.05, indicating that the differences between what we expected and what we observed could have been generated by chance.
For the first burnsi × pipiens cross ($\underline{BB^+} \times B^+B^+$), we would expect a phenotypic ratio of 1:1.

Phenotype	Observed (O)	Expected (E)	$(O{-}E)^2/E$ or (χ^2)
burnsi	23	28	0.89
pipiens	33	28	0.89
Total	56	56	1.78

A chi-square test comparing observed and expected values yields $\chi^2 = 1.78$, df $= 1$, $P > 0.05$. For the second burnsi $\times$ pipiens ($BB^+ \times B^+B^+$) cross, we would expect a phenotypic ratio of 1:1. A chi-square test of the fit of the observed numbers with those expected with a 1:1 ratio yields $\chi^2 = 0.46$, df $= 1$, $P > 0.05$. Thus, all three crosses are consistent with the predication that burnsi is dominant over pipiens.

Phenotype	Observed (O)	Expected (E)	(O–E)^2/E or (χ^2)
burnsi	196	203	0.24
pipiens	210	203	0.24
Total	406	406	0.48

*36. In the California poppy, an allele for yellow flowers (C) is dominant over an allele for white flowers (c). At an independently assorting locus, an allele for entire petals (F) is dominant over an allele for fringed petals (f). A plant that is homozygous for yellow and entire petals is crossed with a plant that is white and fringed. A resulting F_1 plant is then crossed with a plant that is white and fringed, and the following progeny are produced: 54 yellow and entire, 58 yellow and fringed, 53 white and entire, and 10 white and fringed.

a. Use a chi-square test to compare the observed numbers with those expected for the cross.

Solution:
Parents: yellow, entire petals ($CC\ FF$) $\times$ white, fringed petals ($cc\ ff$) $\rightarrow F_1$ ($Cc\ Ff$)
For the cross of a heterozygous F_1 plant ($Cc\ Ff$) with a homozygous recessive plant ($cc\ ff$), we would expect a phenotypic ratio of 1:1:1:1.

Phenotype	Observed (O)	Expected (E)	(O – E)^2/E or (χ^2)
Yellow, entire	54	43.75	2.40
Yellow, fringed	58	43.75	4.64
White, entire	53	43.75	1.96
White, fringed	10	43.75	26.0
Total	175	175	35

A chi-square test comparing the fit of the observed data with the expected 1:1:1:1 ratio yields a chi-square value of 35 with df $= 3$ and $P < 0.005$.

b. What conclusion can you make from the results of the chi-square test?

Solution:
From the chi-square value, it is unlikely that chance produced the differences between the observed and the expected ratio, indicating that the progeny are not in a 1:1:1:1 ratio.

c. Suggest an explanation for the results.

Solution:
The number of plants with the $cc\ ff$ genotype is much less than expected. The $cc\ ff$ genotype is possibly sublethal. Poppies with this genotype are less viable than other genotypes.

Section 3.5

37.   [Data Analysis Problem] Many studies have suggested a strong genetic predisposition to migraine headaches, but the mode of inheritance is not clear. L. Russo and colleagues examined migraine headaches in several families, two of which are shown in the following pedigree (L. Russo et al. 2005. *American Journal of Human Genetics* 76:327–333). What is the most likely mode of inheritance for migraine headaches in these families? Explain your reasoning.

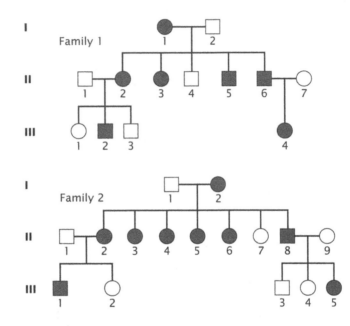

Solution:
In both families, the trait is most likely dominant because it does not skip generations, and affected individuals have one affected parent. In family 2, it is not X-linked because the affected male II-8 has an unaffected daughter. For X-linked loci, an affected male would transmit the trait to all his daughters. It could be either X-linked or autosomal in family 1.

38.   For each of the following pedigrees, give the most likely mode of inheritance, assuming that the trait is rare. Carefully explain your reasoning.

    a.

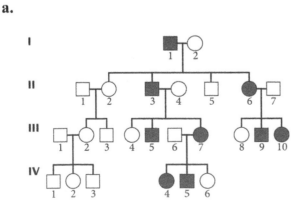

Solution:
Autosomal dominant. The trait must be autosomal because affected males pass on the trait to both sons and daughters. It is dominant because it does not skip generations, all affected individuals have affected parents, and it is extremely unlikely that multiple unrelated individuals mating into the pedigree would be carriers of a rare trait.

b.

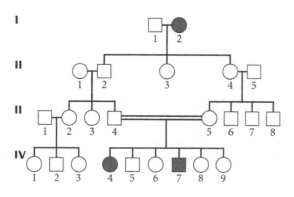

Solution:
Autosomal recessive. Unaffected parents produced affected progeny, so the trait is recessive. The affected daughter must have inherited recessive alleles from both her unaffected parents, so it must be autosomal. If it were X-linked, her father would have shown the trait.

39.   [Data Analysis Problem] Ectrodactyly is a rare condition in which the fingers are absent and the hand is split. This condition is usually inherited as an autosomal dominant trait. Ademar Freire-Maia reported the appearance of ectrodactyly in a family in São Paulo, Brazil, whose pedigree is shown here. Is this pedigree consistent with autosomal dominant inheritance? If not, what mode of inheritance is most likely? Explain your reasoning.

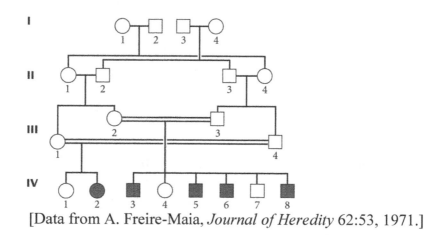

[Data from A. Freire-Maia, *Journal of Heredity* 62:53, 1971.]

Solution:
This pedigree shows autosomal recessive inheritance, not autosomal dominant inheritance. It cannot be dominant because unaffected individuals have affected children. In generation II,

two brothers married two sisters, so the members of generation III in the two families are as closely related as full siblings. A single recessive allele in one of the members of generation I was inherited by all four members of generation III. The consanguineous matings in generation III then produced children homozygous for the recessive ectrodactyly allele. X-linkage is ruled out because the father of female IV-2 is unaffected; he has to be heterozygous.

CHALLENGE QUESTIONS

Section 3.2

*40. A geneticist discovers an obese mouse in his laboratory colony. He breeds this mouse with a normal mouse. All the F_1 mice from this cross are normal in size. When he crosses two F_1 mice, eight of the F_2 mice are normal in size and two are obese. The geneticist then crosses two of his obese mice, and he finds that all the progeny from this cross are obese. These results lead the geneticist to conclude that obesity in mice results from a recessive allele.
A second geneticist at a different university also discovers an obese mouse in her laboratory colony. She carries out the same crosses as those done by the first geneticist and obtains the same results. She also concludes that obesity in mice results from a recessive allele. One day, the two geneticists meet at a genetics conference, learn of each other's experiments, and decide to exchange mice. They both find that, when they cross two obese mice from the different laboratories, all the offspring are normal; however, when they cross two obese mice from the same laboratory, all the offspring are obese. Explain their results.

Solution:
The alleles for obesity from both laboratories are recessive, but they are located at different gene loci. Essentially, the obese mice from the different laboratories have separate obesity genes that are independent of one another.
Possible genotypes of the obese mice are as follows:

Obese mouse from laboratory 1	$(o_1o_1\ O_2O_2)$
Obese mouse from laboratory 2	$(O_1O_1\ o_2o_2)$
P	$o_1o_1\ O_2O_2 \times O_1O_1\ o_2o_2$
F_1 All normal	$(O_1o_1\ O_2o_2)$

41. Albinism in humans is a recessive trait (see the introduction to Chapter 1). A geneticist studies a series of families in which both parents are normal and at least one child has albinism. The geneticist reasons that both parents in these families must be heterozygotes and that albinism should appear in $\frac{1}{4}$ of the children of these families. To his surprise, the geneticist finds that the frequency of albinism among the children of these families is considerably greater than $\frac{1}{4}$. Can you think of an explanation for the higher-than-expected frequency of albinism among these families?

Solution:

The geneticist has indeed identified parents who are heterozygous for the albinism allele. However, by looking only at parents who have albino children, he is missing parents who are heterozygous and have no albino children. Since most parents are probably to have only a few children, the result is that the frequency of albino children produced by parents with an albino child will be higher than what would be predicted. If he were to consider the offspring of normal heterozygous parents with no albino children, along with the parents who have albino children, the expected frequency of albino offspring would most likely approach ¼.

Chapter Four: Extensions and Modifications of Basic Principles

COMPREHENSION QUESTIONS

Section 4.1

*1.   How does sex determination in the XX-XY system differ from sex determination in the ZZ-ZW system?

Solution:
In the XX-XY system, males are heterogametic and produce gametes with either an X chromosome or a Y chromosome. In the ZZ-ZW system, females are heterogametic and produce gametes with either a Z or a W chromosome.

2.   What is meant by genic sex determination?

Solution:
In organisms that follow this system, there is no cytogenetically recognizable difference in the chromosome contents of males and females. Instead of a sex chromosome that differs between males and females, alleles at one or more loci determine the sex of the individual.

3.   How is sex determined in humans?

Solution:
In humans, the presence of a functional Y chromosome determines maleness. People with XXY and XXXY are phenotypically male. In *Drosophila*, the ratio of X chromosome material to autosomes determines the sex of the individual, regardless of the Y chromosome. Flies with XXY are female, and flies with XO are sterile males.

Section 4.2

4.   What characteristics are exhibited by an X-linked trait?

Solution:
Males show the phenotypes of all X-linked traits, regardless of whether the X-linked allele is normally recessive or dominant. Males inherit X-linked traits from their mothers, pass X-linked traits to their daughters, and through their daughters, to their daughters' descendants but not to their sons or their sons' descendants.

5.   Explain why tortoiseshell cats are almost always female and why they have a patchy distribution of orange and black fur.

Solution:
Tortoiseshell cats have two different alleles of an X-linked gene: X^+ (non-orange, or black) and X^o (orange). The patchy distribution results from X-inactivation during early embryo development. Each cell of the early embryo randomly inactivates one of the two

X chromosomes, and the inactivation is maintained in all of the daughter cells. So each patch of black fur arises from a single embryonic cell that inactivated the X^o, and each patch of orange fur arises from an embryonic cell that inactivated the X^+.

6.   What is a Barr body? How is it related to the Lyon hypothesis?

Solution:
Barr bodies are darkly staining bodies in the nuclei of female mammalian cells. Mary Lyon correctly hypothesized that Barr bodies are inactivated (condensed) X chromosomes. By inactivating all X chromosomes beyond one, female cells achieve dosage compensation for X-linked genes.

Section 4.3

7.   How do incomplete dominance and codominance differ?

Solution:
Incomplete dominance means the phenotype of the heterozygote is intermediate between the phenotypes of the two homozygotes. In codominance, both alleles are expressed and both phenotypes are manifested simultaneously.

8.   What is incomplete penetrance and what causes it?

Solution:
In incomplete penetrance, the expected phenotype of a particular genotype is not expressed. Environmental factors, as well as the effects of other genes, may alter the phenotypic expression of a particular genotype.

Section 4.4

9.   What is gene interaction? What is the difference between an epistatic gene and a hypostatic gene?

Solution:
Gene interaction is the determination of a single trait or phenotype by genes at more than one locus; the effect of one gene on a trait depends on the effects of a different gene located elsewhere in the genome. One type of gene interaction is epistasis. The alleles at the epistatic gene mask or repress the effects of alleles at another gene. The gene whose alleles are masked or repressed is called the hypostatic gene.

10.   What is a complementation test and what is it used for?

Solution:
A complementation test is used to determine whether two different recessive mutations are at the same locus (are allelic) or at different loci. The two mutations are introduced into the

same individual organism by crossing homozygotes for each of the mutants. If the progeny shows a mutant phenotype, then the mutations are allelic (at the same locus). If the progeny shows a wild-type (dominant) phenotype, then the mutations are at different loci and are said to complement each other because each of the mutant parents can supply a functional copy (or dominant allele) of the gene mutated in the other parent.

Section 4.5

11.   What characteristics are exhibited by a cytoplasmically inherited trait?

Solution:
Cytoplasmically inherited traits are encoded by genes in the cytoplasm. Because the cytoplasm is usually inherited from a single (most often the female) parent, reciprocal crosses do not show the same results. Cytoplasmically inherited traits often show great variability because different egg cells (female gametes) may have differing proportions of cytoplasmic alleles owing to the random sorting of mitochondria (or chloroplasts in plants).

APPLICATION QUESTIONS AND PROBLEMS

Section 4.1

12.   If nondisjunction of the sex chromosomes takes place in meiosis I in the male in **Figure 4.5**, what sexual phenotypes and proportions of offspring will be produced?

Solution:
Nondisjunction in the male will produce ½ sperm with XY and ½ sperm with O (no sex chromosomes). Normal separation of the sex chromosomes in the female will produce all eggs with a single X chromosome. These gametes will combine to produce ½ XXY (Klinefelter syndrome males) and ½ XO (Turner syndrome females).

13.   What will be the phenotypic sex of a human with the following gene or chromosomes or both?

a. XY with the *SRY* gene deleted

Solution:
Female

b. XY with the *SRY* gene located on an autosome

Solution:
Male

c. XX with a copy of the *SRY* gene on an autosome

Solution:
Male

d. XO with a copy of the *SRY* gene on an autosome

Solution:
Male

e. XXY with the *SRY* gene deleted

Solution:
Female

f. XXYY with one copy of the *SRY* gene deleted

Solution:
Male
In humans, a single functional copy of the *SRY* gene, normally located on the Y chromosome, determines phenotypic maleness by causing gonads to differentiate into testes. In the absence of a functional *SRY* gene, gonads differentiate into ovaries and the individual is phenotypically female.

Section 4.2

*14. Joe has classic hemophilia, an X-linked recessive disease. Could Joe have inherited the gene for this disease from the following persons?

Solution:

	Yes	**No**
a. His mother's mother	X	
b. His mother's father	X	
c. His father's mother		X
d. His father's father		X

X-linked traits are passed on from mother to son. Therefore, Joe must have inherited the hemophilia trait from his mother. His mother could have inherited the trait from either her mother (**a**) or her father (**b**). Because Joe could not have inherited the trait from his father (Joe inherited the Y chromosome from his father), he could not have inherited hemophilia from either (**c**) or (**d**).

*15.   In *Drosophila*, yellow body color is due to an X-linked gene that is recessive to the gene for gray body color.

   a. A homozygous gray female is crossed with a yellow male. The F_1 are intercrossed to produce F_2. Give the genotypes and phenotypes, along with the expected proportions, of the F_1 and F_2 progeny.

   Solution:
   We will use X^+ as the symbol for the dominant gray body color and X^y for the recessive yellow body color. The homozygous gray female parent is thus X^+X^+, and the yellow male parent is X^yY.

   Male progeny always inherit the Y chromosome from the male parent and either of the two X chromosomes from the female parent. Female progeny always inherit the X chromosome from the male parent and either of the two X chromosomes from the female parent.

   F_1 males inherit the Y chromosome from their father and X^+ from their mother; hence, their genotype is X^+Y, and they have gray bodies. F_1 females inherit X^y from their father and X^+ from their mother; hence, they are X^+X^y and also have gray bodies.

   When the F_1 progeny are intercrossed, the F_2 males again inherit the Y from the F_1 male, and they inherit either X^+ or X^y from their mother. Therefore, we should get ½ X^+Y (gray body) and ½ X^yY (yellow body). The F_2 females will all inherit the X^+ from their father and either X^+ or X^y from their mother. Therefore, we should get ½ X^+X^+ and ½ X^+X^y (all gray body).

   In summary:
   P X^+X^+ (gray female) × X^yY (yellow male)
   F_1 ½ X^+Y (gray males)
            ½ X^+X^y (gray females)
   F_2 ¼ X^+Y (gray males)
            ¼ X^yY (yellow males)
            ¼ X^+X^y (gray females)
            ¼ X^+X^+ (gray females)

   The net F_2 phenotypic ratios are ½ gray females, ¼ gray males, and ¼ yellow males. The F_2 progeny can also be predicted using a Punnett square.

X^+	Y
X^+X^+ gray females	X^+Y gray males
X^+X^y gray females	X^yY yellow males

   b. A yellow female is crossed with a gray male. The F_1 are intercrossed to produce the F_2. Give the genotypes and phenotypes, along with the expected proportions, of the F_1 and F_2 progeny.

Solution:

The yellow female must be homozygous X^yX^y because yellow is recessive, and the gray male, having only one X chromosome, must be X^+Y. The F_1 male progeny are all X^yY (yellow) and the F_1 females are all X^+X^y (heterozygous gray).

P X^yX^y (yellow female) × X^+Y (gray male)

F_1 ½ X^yY (yellow males)

       ½ X^+X^y (gray females)

F_2 ¼ X^+Y (gray males)

       ¼ X^yY (yellow males)

       ¼ X^+X^y (gray females)

       ¼ X^yX^y (yellow females)

16.   Coat color in cats is determined by genes at several different loci. At one locus on the X chromosome, one allele (X^+) encodes black fur; another allele (X^o) encodes orange fur. Females can be black (X^+X^+), orange (X^oX^o) or a mixture of orange and black called tortoiseshell (X^+X^o). Males are either black (X^+Y) or orange (X^oY). Bill has a female tortoiseshell cat named Patches. One night Patches escapes from Bill's house, spends the night out, and mates with a stray male. Patches later gives birth to the following kittens: one orange male, one black male, two tortoiseshell females, and one orange female. Give the genotypes of Patches, her kittens, and the stray male with which Patches mated.

Solution:

Patches: X^oX^+. Male: X^oY. Kittens: X^oY, X^+Y, X^+X^o, X^oX^o.

17.   Red–green color blindness in humans is due to an X-linked recessive gene. Both John and Cathy have normal color vision. After 10 years of marriage to John, Cathy gave birth to a color-blind daughter. John filed for divorce, claiming he is not the father of the child. Is John justified in his claim of nonpaternity? Explain why. If Cathy had given birth to a color-blind son, would John be justified in claiming nonpaternity?

Solution:

Because color blindness is a recessive trait, the color-blind daughter must be homozygous recessive. Because color blindness is X-linked, John has grounds for suspicion.

Normally, their daughter would have inherited John's X chromosome. Because John is not color blind, he could not have transmitted an X chromosome with a color-blind allele to his daughter.

A remote alternative possibility is that the daughter is XO, having inherited a recessive color-blind allele from her mother and no sex chromosome from her father. In that case, the daughter would have Turner syndrome. A new X-linked color-blind mutation is also possible, albeit even less likely.

If Cathy had a color-blind son, then John would have no grounds for suspicion. The son would have inherited John's Y chromosome and the color-blind X chromosome from Cathy.

18.   [Data Analysis Problem] The following pedigree illustrates the inheritance of Nance–Horan syndrome, a rare genetic condition in which affected persons have cataracts and abnormally shaped teeth.

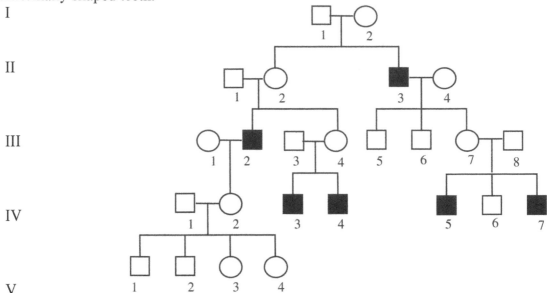

(Pedigree after D. Stambolian, R. A. Lewis, K. Buetow, A. Bond, and R. Nussbaum. *American Journal of Human Genetics* 47:15, 1990.)

a.   On the basis of this pedigree, what do you think is the most likely mode of inheritance for Nance–Horan syndrome?

Solution:
X-linked recessive. Only males have the condition, and unaffected female carriers have affected sons.

b.   If couple III-7 and III-8 have another child, what is the probability that the child will have Nance–Horan syndrome?

Solution:
The probability is ¼. The female III-7 is a carrier, so there is a ½ probability that the child will inherit her X chromosome with the Nance–Horan allele and another ½ probability that the child will be a boy.

c.   If III-2 and III-7 were to mate, what is the probability that one of their children would have Nance–Horan syndrome?

Solution:
The probability is ½ because half the boys will inherit the Nance–Horan allele from the III-7 carrier female. All the girls will inherit one Nance–Horan allele from the III-2 affected male, and half of them will get a second Nance–Horan allele from the III-7 female, so half the girls will also have Nance–Horan syndrome.

*19.   Bob has XXY chromosomes (Klinefelter syndrome) and is color blind. His mother and father have normal color vision, but his maternal grandfather is color blind. Assume that Bob's chromosome abnormality arose from nondisjunction in meiosis. In which parent and in which meiotic division did nondisjunction take place? Assume that no crossing over has taken place. Explain your answer.

Solution:
Because Bob must have inherited the Y chromosome from his father and his father has normal color vision, a nondisjunction event in the paternal lineage cannot account for Bob's genotype. Bob's mother must be heterozygous X^+X^c because she has normal color vision, and she must have inherited a color-blind X chromosome from her color-blind father. For Bob to inherit two color-blind X chromosomes from his mother, the egg must have arisen from a nondisjunction in meiosis II. In meiosis I, the homologous X chromosomes separate, and so one cell has the X^+ and the other has X^c. The failure of sister chromatids to separate in meiosis II would then result in an egg with two copies of X^c.

20.   The Talmud, an ancient book of Jewish civil and religious laws, states that if a woman bears two sons who die of bleeding after circumcision (removal of the foreskin from the penis), any additional sons that she has should not be circumcised. (The bleeding is most likely due to the X-linked disorder hemophilia.) Furthermore, the Talmud states that the sons of her sisters must not be circumcised, whereas the sons of her brothers should. Is this religious law consistent with sound genetic principles? Explain your answer.

Solution:
Yes. If a woman has a son with hemophilia, then she is a carrier. Any of her sons have a 50% chance of having hemophilia. Her sisters may also be carriers. Her brothers, if they do not themselves have hemophilia (because they survived circumcision, they most likely do not have hemophilia), cannot be carriers, and therefore there is no risk of passing hemophilia on to their children.

21.   [Data Analysis Problem] Craniofrontonasal syndrome (CFNS) is a birth defect in which premature fusion of the cranial sutures leads to abnormal head shape, widely spaced eyes, nasal clefts, and various other skeletal abnormalities. George Feldman and his colleagues looked at several families in which CFNS occurred and recorded the results shown in the following table (G. J. Feldman. 1997. *Human Molecular Genetics* 6:1937–1941).

Family number	Parents		Offspring			
	Father	Mother	Normal male	Female	CFNS male	Female
1	normal	CFNS	1	0	2	1
5	normal	CFNS	0	2	1	2
6	normal	CFNS	0	0	1	2
8	normal	CFNS	1	1	1	0
10a	CFNS	normal	3	0	0	2
10b	normal	CFNS	1	1	2	0
12	CFNS	normal	0	0	0	1
13a	normal	CFNS	0	1	2	1
13b	CFNS	normal	0	0	0	2
7b	CFNS	normal	0	0	0	2

a. On the basis of the families given, what is the most likely mode of inheritance for CFNS?

Solution:
Children with CFNS are born in families where one parent has CFNS. This is most likely a dominant trait. Moreover, we see that in families where the father has CFNS and the mother does not, CFNS is transmitted only to girls, not to boys. If the mother has CFNS, then both boys and girls get CFNS. These data are consistent with an X-linked dominant mode of inheritance for CFNS.

b. Give the most likely genotypes of the parents in families 1 and 10a.

Solution:
Family 1: normal father (X^+Y) × CFNS mother (X^+X^c)
Family 10a: CFNS father (X^cY) × normal mother (X^+X^+)

22. How many Barr bodies would you expect to see in a human cell containing the following chromosomes?

Solution:
a. XX—1 Barr body
b. XY—0
c. XO—0
d. XXY—1
e. XXYY—1
f. XXXY—2
g. XYY—0
h. XXX—2
i. XXXX—3

Human cells inactivate all X chromosomes beyond one. The Y chromosome has no effect on X-inactivation.

23.   What is the most likely sex and genotype of the cat shown in **Figure 4.12**?

Solution:
The cat has the tortoiseshell phenotype, which is produced by being heterozygous for black and orange alleles at the X-linked color locus (X^+X^o). Because the genes for black and orange are X-linked, tortoiseshell cats are female; males normally only have a single X chromosome and therefore cannot be heterozygous.

24.   Red–green color blindness is an X-linked recessive trait in humans. Polydactyly (extra fingers and toes) is an autosomal dominant trait. Martha has normal fingers and toes and normal color vision. Her mother is normal in all respects, but her father is color blind and polydactylous. Bill is color blind and polydactylous. His mother has normal color vision and normal fingers and toes. If Bill and Martha marry, what types and proportions of children can they produce?

Solution:
The first step is to deduce the genotypes of Martha and Bill. Because the two traits are independent, we can deal with just one trait at a time.

Starting with the X-linked color-blind trait, Bill must be X^cY because he is colorblind. Bill's mother must be a carrier (X^+X^c). Martha must be X^+X^c, a carrier for color blindness because her father is color blind (X^cY).

For polydactyly, Bill must be Dd (D denotes the dominant polydactyly allele). Because his mother has normal fingers (dd), he cannot be homozygous DD. Martha, with normal fingers, must be dd.

If Martha (dd, X^+X^c) marries Bill (Dd, X^cY), then we can predict the types and probability ratios of children they could produce.

For polydactyly, ½ of children will be polydactylous and ½ will have normal fingers.

For color blindness, ¼ of children will be color-blind girls, ¼ will be girls with normal vision but carrying the color blindness allele, ¼ will be color-blind boys, and ¼ will be boys with normal vision.

Combining both traits, then:
    1/8 color-blind girls with normal fingers
    1/8 color-blind girls with polydactyly
    1/8 girls with normal vision and normal fingers
    1/8 girls with normal vision and polydactyly
    1/8 color-blind boys with normal fingers
    1/8 color-blind boys with polydactyly
    1/8 boys with normal vision and normal fingers
    1/8 boys with normal vision and polydactyly
    This analysis can also be carried out with a Punnett square.

*25.   Miniature wings in *Drosophila* result from an X-linked gene (X^m) that is recessive to an allele for long wings (X^+). Sepia eyes are produced by an autosomal gene (s) that is recessive to an allele for red eyes (s^+).

   a.   A female fly that has miniature wings and sepia eyes is crossed with a male that has normal wings and is homozygous for red eyes. The F$_1$ are intercrossed to produce the F$_2$. Give the phenotypes and their proportions expected in the F$_1$ and F$_2$ flies from this cross.

Solution:
The female parent (miniature wings, sepia eyes) must be $X^m X^m$, ss.
The male parent (normal wings, homozygous red eyes) is X^+Y, s^+s^+.

    The F$_1$ males are $X^m Y$, s^+s (miniature wings, red eyes)
    The F$_1$ females are X^+X^m, s^+s (long wings, red eyes)

The proportions expected in the F$_2$ can be obtained by breaking the cross into two separate crosses, one for each locus, and then combining the probabilities with the branch method.

First, let's determine the outcome for the locus that codes for eye color.

$s^+s \times s^+s$

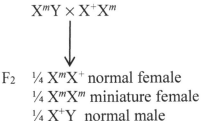

¾ s^+_ red eyes
¼ ss sepia eyes

Next, let's consider the locus that determines wing length.

    $X^m Y \times X^+ X^m$

F$_2$ ¼ $X^m X^+$ normal female
      ¼ $X^m X^m$ miniature female
      ¼ $X^+ Y$ normal male
      ¼ $X^m Y$ miniature male

Now, we can use the branch method to combine the probability of obtaining both characteristics together in the F$_2$.

Eye color	Wing shape and sex	F$_2$ phenotype	Probability
Red (¾)	normal female → (¼)	Red eyed, normal winged female	¾ × ¼ = 3/16
	miniature female→ (¼)	Red eyed, miniature winged female	¾ × ¼ = 3/16
	normal male → (¼)	Red eyed, normal winged male	¾ × ¼ = 3/16
	miniature male→ (¼)	Red eyed, miniature winged male	¾ × ¼ = 3/16
Sepia (¼)	normal female→ (¼)	Sepia eyed, normal winged female	¼ × ¼ = 1/16
	miniature female→ (¼)	Sepia eyed, miniature winged female	¼ × ¼ = 1/16
	normal male → (¼)	Sepia eyed, normal winged male	¼ × ¼ = 1/16
	miniature male→ (¼)	Sepia eyed, miniature winged male	¼ × ¼ = 1/16

b. A female fly that is homozygous for normal wings and has sepia eyes is crossed with a male that has miniature wings and is homozygous for red eyes. The F$_1$ are intercrossed to produce the F$_2$. Give the phenotypes and their proportions expected in the F$_1$ and F$_2$ flies from this cross.

Solution:

The female parent (homozygous for normal wings, sepia eyes) is X^+X^+, ss

The male parent (miniature wings, homozygous red eyes) is X^mY, s^+s^+

The F$_1$ males are X^+Y, s^+s (normal wings, red eyes)

The F$_1$ females are X^+X^m, s^+s (normal wings, red eyes)

At the eye color locus, we expect the following proportions in the F$_2$.

$s^+s \times s^+s$

$\downarrow$

¾ s^+_ red eyes

¼ ss sepia eyes

At the locus that determines wing length, we expect the following proportions in the F$_2$.

$X^+Y \times X^+X^m$

$\downarrow$

F$_2$ ¼ X^+X^+ normal female   ⎫
   ¼ X^+X^m normal female   ⎬ ½ normal females
   ¼ X^+Y normal male   ⎭
   ¼ X^mY miniature male

Now, we can apply the branch method to combine the probability of obtaining both characteristics together in the F2.

Eye color	Wing shape and sex	F2 phenotype	Probability
Red (¾)	normal female → (½)	Red eyed, long winged female	¾ × ½ = 3/8
	normal male → (¼)	Red eyed, long winged male	¾ × ¼ = 3/16
	miniature male → (¼)	Red eyed, miniature winged male	¾ × ¼ = 3/16
Sepia (¼)	normal female → (½)	Sepia eyed, long winged female	¼ × ½ = 1/8
	normal male → (¼)	Sepia eyed, long winged male	¼ × ¼ = 1/16
	miniature male → (¼)	Sepia eyed, miniature winged male	¼ × ¼ = 1/16

Section 4.3

*26. Palomino horses have a golden yellow coat, chestnut horses have a brown coat, and cremello horses have a coat that is almost white. A series of crosses between the three different types of horses produced the following offspring:

Cross	Offspring
palomino × palomino	13 palomino, 6 chestnut, 5 cremello
chestnut × chestnut	16 chestnut
cremello × cremello	13 cremello
palomino × chestnut	8 palomino, 9 chestnut
palomino × cremello	11 palomino, 11 cremello
chestnut × cremello	23 palomino

a. Explain the inheritance of the palomino, chestnut, and cremello phenotypes in horses.

Solution:
The results of the crosses indicate that cremello and chestnut are pure-breeding traits (homozygous). Palomino is a hybrid trait (heterozygous) that produces a 2:1:1 ratio when palominos are crossed with each other. The simplest hypothesis consistent with these results is incomplete dominance, with palomino as the phenotype of the heterozygotes resulting from chestnuts crossed with cremellos.

b. Assign symbols for the alleles that determine these phenotypes, and list the genotypes of all parents and offspring given in the preceding table.

Solution:
Let C^B = chestnut, C^W = cremello, $C^B C^W$ = palomino.

Cross		**Offspring**

Cross	Offspring
palomino × palomino $C^BC^W \times C^BC^W$	13 palomino, 6 chestnut, 5 cremello C^BC^W C^BC^B C^WC^W
chestnut × chestnut $C^BC^B \times C^BC^B$	16 chestnut C^BC^B
cremello × cremello $C^WC^W \times C^WC^W$	13 cremello C^WC^W
palomino × chestnut $C^BC^W \times C^BC^B$	8 palomino, 9 chestnut C^BC^W C^BC^B
palomino × cremello $C^BC^W \times C^WC^W$	11 palomino, 11 cremello C^BC^W C^WC^W
chestnut × cremello $C^BC^B \times C^WC^W$	23 palomino C^BC^W

27. The L^M and L^N alleles at the MN blood-group locus exhibit codominance. Give the expected genotypes and phenotypes and their ratios in progeny resulting from the following crosses:

a. $L^M L^M \times L^M L^M$

Solution:
½ L^ML^M (type M), ½ L^ML^N (type MN)

b. $L^NL^N \times L^NL^N$

Solution:
All L^NL^N (type N)

c. $L^ML^N \times L^ML^N$

Solution:
½ L^ML^N (type MN), ¼ L^ML^M (type M), ¼ L^NL^N (type N)

d. $L^ML^N \times L^NL^N$

Solution:
½ L^ML^N (type MN), ½ L^NL^N (type N)

e. $L^ML^M \times L^NL^N$

Solution:
All L^ML^N (type MN)

*28. Assume that long earlobes in humans are an autosomal dominant trait that exhibits 30% penetrance. A person who is heterozygous for long earlobes mates with a person who is homozygous for normal earlobes. What is the probability that their first child will have long earlobes?

Solution:
To have long earlobes, the child must inherit the dominant allele and also express it. The probability of inheriting the dominant allele is 50%; the probability of expressing it is 30%. The combined probability of both is 0.5(0.3) = 0.15, or 15%.

*29.   When a Chinese hamster with white spots is crossed with another hamster that has no spots, approximately ½ of the offspring have white spots and ½ have no spots. When two hamsters with white spots are crossed, ⅔ of the offspring possess white spots and ⅓ have no spots.

a.   What is the genetic basis of white spotting in Chinese hamsters?

Solution:
The 2:1 ratio in the progeny of two spotted hamsters suggests lethality, and the 1:1 ratio in the progeny of a spotted hamster and a hamster without spots indicates that spotted is a heterozygous phenotype. If S and s represent the locus for white spotting, spotted hamsters are Ss and solid-colored hamsters are ss. One-quarter of the zygotes expected from a mating of two spotted hamsters is SS, embryonic lethal, and missing from the progeny, resulting in the 2:1 ratio of spotted to solid progeny.

b.   How might you go about producing Chinese hamsters that breed true for white spotting?

Solution:
Because spotting is a heterozygous phenotype, it should not be possible to obtain Chinese hamsters that breed true for spotting.

30.   [Data Analysis Problem] In the early 1900s, Lucien Cuénot, a French scientist working at the University of Nancy, studied the genetic basis of yellow coat color in mice. He carried out a number of crosses between two yellow mice and obtained what he thought was a 3:1 ratio of yellow to gray mice in the progeny. The following table gives Cuénot's actual results, along with the results of a much larger series of crosses carried out by William Castle and Clarence Little (W.E. Castle and C. C. Little. 1910. *Science* 32:868–870).

Investigators	Yellow progeny	Nonyellow progeny	Total progeny
Cuénot	263	100	363
Castle and Little	800	435	1235
Both combined	1063	535	1598

a.   Using a chi-square test, determine whether Cuénot's results are significantly different from the 3:1 ratio that he thought he observed. Are they different from a 2:1 ratio?

Solution:
Testing Cuénot's data for a 3:1 ratio:

	Obs	Expected (3:1)	O − E	(O − E)²/E
Yellow	263	272.25	−9.25	0.314
Non-yellow	100	90.75	9.25	0.943
Total	363	363		$1.257 = \chi^2$

d.f. = 2 − 1 = 1; $0.1 < p < 0.5$
Cannot reject hypothesis of 3:1 ratio.

Now test for 2:1 ratio:

	Obs	Expected (2:1)	O − E	(O − E)²/E
Yellow	263	242	21	1.82
Non-yellow	100	121	−21	3.64
Total	363	363		$5.46 = \chi^2$

d.f. = 1; $p < 0.025$
The observations are inconsistent with a 2:1 ratio.

c. Determine whether Castle and Little's results are significantly different from a 3:1 ratio. Are they different from a 2:1 ratio?

Solution:

	Obs	Expected (3:1)	O − E	(O − E)²/E
Yellow	800	926.25	−126.25	17.2
Non-yellow	435	308.75	126.25	51.6
Total	1235	1235		$68.8 = \chi^2$

d.f. = 1; $p \ll 0.005$
Reject 3:1 ratio.

	Obs	Expected (2:1)	O − E	(O − E)²/E
Yellow	800	823.3	−23.3	0.66
Non-yellow	435	411.7	23.3	1.32
Total	1235	1235		$1.98 = \chi^2$

d.f. = 1; $0.1 < p < 0.5$
Cannot reject 2:1 ratio.

d. Combine the Castle and Little results with those of Cuénot and determine whether they are significantly different from a 3:1 ratio and a 2:1 ratio.

Solution:

	Obs	Expected (3:1)	O − E	(O − E)²/E
Yellow	1063	1198.5	−135.5	15.3
Non-yellow	535	399.5	135.5	46.0
Total	1598	1598		$61.3 = \chi^2$

d.f. = 1; $p \ll 0.005$
Reject 3:1 ratio.

	Obs	Expected (2:1)	O − E	(O − E)²/E
Yellow	1063	1065.3	−2.3	0.005
Non-yellow	535	532.7	2.3	0.010
Total	1598	1598		$0.015 = \chi^2$

d.f. = 1; $0.9 < p < 0.975$
Cannot reject 2:1 ratio.

e. Offer an explanation for the different ratios obtained by Cuénot and by Castle and Little.

Solution:
Cuénot had far smaller numbers of progeny, so his ratios are more susceptible to error from chance deviation. Indeed, only a slight shift in numbers of progeny would make Cuénot's data compatible with a 2:1 ratio as well as a 3:1 ratio. Investigator bias may also have played a role, based on the expectation of a 3:1 ratio.

31. In rabbits, an allelic series helps to determine coat color: C (full color), c^{ch} (chinchilla, gray color), c^h (Himalayan, white with black extremities), and c (albino, all white). The C allele is dominant over all others, c^{ch} is dominant over c^h and c, c^h is dominant over c, and c is recessive to all the other alleles. This dominance hierarchy can be summarized as $C > c^{ch} > c^h > c$. The rabbits in the following list are crossed and produce the progeny shown. Give the genotypes of the parents for each cross.

Phenotypes of parents	**Phenotypes of offspring**
a. full color × albino	½ full color, ½ albino

Solution:
$Cc \times cc$. 1:1 phenotypic ratios in the progeny result from a cross of a heterozygote with a homozygous recessive. Because albino is recessive to all other alleles, the full-color parent must have an albino allele, and the albino parent must be homozygous for the albino allele.

b. Himalayan × albino   ½ Himalayan, ½ albino

Solution:
$c^h c \times cc$. Again, the 1:1 ratio of the progeny indicates the parents must be a heterozygote and a homozygous recessive.

c. full color × albino   ½ full color, ½ chinchilla

Solution:
$Cc^{ch} \times cc$. This time, we get a 1:1 ratio, but we have chinchilla progeny instead of albino. Therefore, the heterozygous full-color parent must have a chinchilla allele as well as a dominant full-color allele. The albino parent has to be homozygous albino because albino is recessive to all other alleles.

d. full color × Himalayan ½ full color, ¼ Himalayan, ¼ albino

Solution:
$Cc \times c^h c$. The 1:2:1 ratio in the progeny indicates that both parents are heterozygotes. Both must have an albino allele because the albino progeny must have inherited an albino allele from each parent.

e. full color × full color ¾ full color, ¼ albino

Solution:
$Cc \times Cc$. The 3:1 ratio indicates that both parents are heterozygous. Both parents must have an albino allele for albino progeny to result.

*32. In this chapter, we considered Joan Barry's paternity suit against Charlie Chaplin and how, on the basis of blood types, Chaplin could not have been the father of her child.

a. What blood types are possible for the father of Barry's child?

Solution:
Because Barry's child inherited an I^B allele from the father, the father could have been B or AB.

b. If Chaplin had possessed one of these blood types, would that prove that he fathered Barry's child?

Solution:
No. Many other men have these blood types. The results would have meant only that Chaplin cannot be eliminated as a possible father of the child.

Section 4.4

*33. In chickens, comb shape is determined by alleles at two loci (R, r and P, p). A walnut comb is produced when at least one dominant allele R is present at one locus and at least one dominant allele P is present at a second locus (genotype $R\_ P\_$). A rose comb is produced when at least one dominant allele is present at the first locus and two recessive alleles are present at the second locus (genotype $R\_ pp$). A pea comb is produced when two recessive alleles are present at the first locus and at least one dominant allele is present at the second (genotype $rr P\_$). If two recessive alleles are present at the first locus and at the second locus ($rr pp$), a single comb is produced. Progeny with what types of combs and in what proportions will result from the following crosses?

a. $RR\ PP \times rr\ pp$

Solution:
All walnut ($Rr\ Pp$)

b. *Rr Pp × rr pp*

Solution:
¼ walnut (*Rr Pp*), ¼ rose (*Rr pp*), ¼ pea (*rr* Pp), ¼ single (*rr pp*)

c. *Rr Pp × Rr Pp*

Solution:
$\frac{9}{16}$ walnut (*R_ P_*), $\frac{3}{16}$ rose (*R_pp*), $\frac{3}{16}$ pea (*rr P_*), $\frac{1}{16}$ single (*rr pp*)

d. *Rr pp × Rr pp*

Solution:
¾ rose (*R_pp*), ¼ single (*rr pp*)

e. *Rr pp × rr Pp*

Solution:
¼ walnut (*Rr Pp*), ¼ rose (*Rr pp*), ¼ pea (*rr Pp*), ¼ single (*rr pp*)

f. *Rr pp × rr pp*

Solution:
½ rose (*Rr pp*), ½ single (*rr pp*)

34.   [Data Analysis Problem] Tatuo Aida investigated the genetic basis of color variation in the medaka (*Aplocheilus latipes*), a small fish found in Japan (T. Aida. 1921. *Genetics* 6:554–573). Aida found that genes at two loci (*B, b* and *R, r*) determine the color of the fish: fish with a dominant allele at both loci (*B_ R_*) are brown, fish with a dominant allele at the *B* locus only (*B_ rr*) are blue, fish with a dominant allele at the *R* locus only (*bb R_*) are red, and fish with recessive alleles at both loci (*bb rr*) are white. Aida crossed a homozygous brown fish with a homozygous white fish. He then backcrossed the F₁ with the homozygous white parent and obtained 228 brown fish, 230 blue fish, 237 red fish, and 222 white fish.

   a.   Give the genotypes of the backcross progeny.

   Solution:
   Each of the backcross progeny received recessive alleles *b* and *r*. Their phenotype is therefore determined by the alleles received from the other parent:
   brown fish are *Bb Rr*; blue fish are *Bb rr*; red fish are *bb Rr*; and white fish are *bb rr*.

   b.   Use a chi-square test to compare the observed numbers of backcross progeny with the number expected. What conclusion can you make from your chi-square results?

   Solution:
   We expect a 1:1:1:1 ratio of the four phenotypes.

	Observed	Expected	O – E	$(O – E)^2/E$
Brown	228	229.25	–1.25	0.007
Blue	230	229.25	0.75	0.002
Red	237	229.25	7.75	0.262
White	222	229.25	–7.25	0.229
Total	917	917		$0.5 = \chi^2$

d.f. $= 4 – 1 = 3$; $.9 < p < 0.975$; we cannot reject the hypothesis.

c. What results would you expect for a cross between a homozygous red fish and a white fish?

Solution:
The homozygous red fish would be *bb RR*, crossed to *bb rr*. All progeny would be *bb Rr*, or red fish.

d. What results would you expect if you crossed a homozygous red fish with a homozygous blue fish and then backcrossed the F_1 with a homozygous red parental fish?

Solution:
Homozygous red fish *bb RR* × homozygous blue fish *BB rr*
F_1 will be all brown: *Bb Rr* backcrossed *to bb RR* (homozygous red parent).
Backcross progeny will be in equal proportions *Bb RR* (brown); *Bb Rr* (brown); *bb RR* (red); and *bb Rr* (red). Overall, ½ brown and ½ red.

35. [Data Analysis Problem] E. W. Lindstrom crossed two corn plants with green seedlings and obtained the following progeny: 3583 green seedlings, 853 virescent-white seedlings, and 260 yellow seedlings (E. W. Lindstrom. 1921. *Genetics* 6:91–110).

a. Give the genotypes for the green, virescent-white, and yellow progeny.

Solution:
There are 4696 total progeny. Green appears dominant. The ratios at first glance don't fit any type of incomplete dominance for a single locus, so we hypothesize multiple loci with gene interactions. The simplest case is two loci, so we look for a fit to a ratio based on $^1/_{16}$ of the total: 293.5. Quick computation with a calculator shows that these numbers are close to a 12:3:1 ratio of green:virescent-white:yellow, a modified 9:3:3:1 ratio. Let's define *G* and *g* for one locus, and *Y* and *y* for the other locus.

9 *G_ Y_* + 3 *G_ yy* = 12 green
3 *gg Y_* = 3 virescent-white
1 *gg yy* = 1 yellow

b. Explain how color is determined in these seedlings.

Solution:
The green arises when the *G* locus is dominant, regardless of the alleles at the other *Y* locus. Yellow requires that both loci be recessive, and virescent-white arises when the *G* locus is homozygous recessive, and the *Y* locus has a dominant allele.

c. Does epistasis take place among the genes that determine color in the corn seedlings? If so, which gene is epistatic and which is hypostatic?

Solution:
As defined above, the G locus is the epistatic locus. It is an example of dominant epistatis because a dominant allele at this locus masks the effect of the Y locus. The Y locus is hypostatic, and its effect revealed only when the epistatic locus is homozygous recessive.

*36. A summer-squash plant that produces disc-shaped fruit is crossed with a summer-squash plant that produces long fruit. All the F_1 have disc-shaped fruit. When the F_1 are intercrossed, F_2 progeny are produced in the following ratio: $^9/_{16}$ disc-shaped fruit: $^6/_{16}$ spherical fruit; $^1/_{16}$ long fruit. Give the genotypes of the F_2 progeny.

Solution:
The modified dihybrid ratio in the F_2 indicates that two genes interact to determine fruit shape. Let A and B represent the two loci. The F_1 heterozygotes are $Aa\ Bb$.
   The F_2 are:
$^9/_{16}\ A\_\ B\_$ disc-shaped (like F_1)
$^3/_{16}\ A\_\ bb$ spherical
$^3/_{16}\ aa\ B\_$ spherical
$^1/_{16}\ aa\ bb$ long

37.   Some sweet-pea plants have purple flowers and other plants have white flowers. A homozygous variety of sweet pea that has purple flowers is crossed with a homozygous variety that has white flowers. All the F_1 have purple flowers. When these F_1 are self-fertilized, the F_2 appear in a ratio of $^9/_{16}$ purple to $^7/_{16}$ white.

a. Give genotypes for the purple and white flowers in these crosses.

Solution:
The F_2 ratio of 9:7 is a modified dihybrid ratio, indicating two genes interacting. Using A and B as generic gene symbols, we can start with the F_1 heterozygotes:
   F_1 $Aa\ Bb$ purple self-fertilized
   F_2 $^9/_{16}\ A\_\ B\_$ purple (like F_1)
            $^3/_{16}\ A\_\ bb$ white
            $^3/_{16}\ aa\ B\_$ white
            $^1/_{16}\ aa\ bb$ white
Now we see that purple requires dominant alleles for both genes, so the purple parent must have been $AA\ BB$, and the white parent must have been $aa\ bb$ to give all purple F_1.

b. Draw a hypothetical biochemical pathway to explain the production of purple and white flowers in sweet peas.

Solution:
White precursor 1 ⟶ white intermediate 2 ⟶ purple pigment
            Enzyme A   Enzyme B

Section 4.5

*38. The direction of shell coiling in the snail *Lymnaea peregra* (discussed in the introduction to this chapter) results from a genetic maternal effect. An autosomal allele for a right-handed shell (s^+), called dextral, is dominant over the allele for a left-handed shell (s), called sinistral. A pet snail called Martha is sinistral and reproduces only as a female (the snails are hermaphroditic). Indicate which of the following statements are true and which are false. Explain your reasoning in each case.

   a. Martha's genotype *must* be *ss*.

   Solution:
   False. For maternal effect genes, the phenotype of the individual is determined solely by the genotype of the individual's mother. So we know Martha's mother must have been *ss* because Martha is sinistral. If Martha was produced as a result of self-fertilization, then Martha must indeed be *ss*. But if Martha was produced by cross-fertilization, then we cannot know Martha's genotype without more information.

   b. Martha's genotype *cannot* be s^+s^+.

   Solution:
   True. As explained in the answer to part (*a*), Martha's mother is *ss*, so Martha must be either s^+s or *ss*.

   c. All the offspring produced by Martha *must* be sinistral.

   Solution:
   False. Because we do not know Martha's genotype, we cannot yet predict the phenotype of her offspring.

   d. At least some of the offspring produced by Martha *must* be sinistral.

   Solution:
   False. If Martha is s^+s, then all her children will be dextral. If Martha is *ss*, then all her children will be sinistral.

   e. Martha's mother *must* have been sinistral.

   Solution:
   False. Martha's mother's phenotype is determined by the genotype of her mother (Martha's maternal grandmother). We know Martha's mother's genotype must have been *ss*, so her mother's mother had at least one *s* allele. But we cannot know if she was a heterozygote or homozygous *ss*.

   f. All Martha's brothers *must* be sinistral.

   Solution:
   True. Because Martha's mother must have been *ss*, all her progeny must be sinistral.

39.   If the F_2 dextral snails with genotype s^+s in **Figure 4.21** undergo self-fertilization, what phenotypes and proportions are expected to occur in the progeny?

Solution:
All dextral

Section 4.6

40.   Which of the following statements is an example of a phenocopy? Explain your reasoning.

   a.   Phenylketonuria results from a recessive mutation that causes light skin as well as intellectual disability.

   Solution:
   Phenocopy is an environmentally induced phenotype that resembles a phenotype produced by a genotype. Since phenylketonuria has a genetic basis, this is not a phenocopy. One genotype affecting multiple traits is called pleiotropy.

   b.   Human height is influenced by genes at many different loci.

   Solution:
   This is again not an example of phenocopy, but of a continuous characteristic or a quantitative trait.

   c.   Dwarf plants and mottled leaves in tomatoes are caused by separate genes that are linked.

   Solution:
   Linkage of genes is not an example of phenocopy.

   d.   Vestigial wings in *Drosophila* are produced by a recessive mutation. This trait is also produced by high temperature during development.

   Solution:
   This is indeed an example of phenocopy because an environmental factor produces a phenotype that resembles the phenotype generated by a genotype.

   e.   Intelligence in humans is influenced by both genetic and environmental factors.

   Solution:
   As long as there is a significant effect of the underlying genotype, this is not a phenocopy. The expression of many genotypes is indeed influenced by environmental factors.

41.   Match each of the following terms with its correct definition (parts *a* through *h*).

Solution:
_**d**_ phenocopy   _**h**_ genetic maternal effect
_**g**_ pleiotropy   _**f**_ genomic imprinting
_**e**_ polygenic trait   _**b**_ sex-influenced trait
_**a**_ penetrance
_**c**_ sex-limited trait

a. The percentage of individuals with a particular genotype that express the expected phenotype
b. A trait determined by an autosomal gene that is more easily expressed in one sex
c. A trait determined by an autosomal gene that is expressed in only one sex
d. A trait that is determined by an environmental effect and has the same phenotype as a genetically determined trait
e. A trait determined by genes at many loci
f. The expression of a trait affected by the sex of the parent that transmits the gene to the offspring
g. A gene affecting more than one phenotype
h. The influence of the genotype of the maternal parent on the phenotype of the offspring

CHALLENGE QUESTION

Section 4.2

42.   A geneticist discovers a male mouse with greatly enlarged testes in his laboratory colony. He suspects that this trait results from a new mutation that is either Y-linked or autosomal dominant. How could he determine which it is?

Solution:
Because testes are present only in males, enlarged testes could either be a sex-limited autosomal dominant trait or a Y-linked trait. Assuming the male mouse with enlarged testes is fertile, mate it with a normal female. If the trait is autosomal dominant and the parental male is heterozygous, only half the male progeny will have enlarged testes. If the trait is Y-linked, all the male progeny will have enlarged testes.

With either outcome, however, the results from this first cross will not be conclusive. If all the male progeny do express the trait, the trait may still be autosomal dominant if the parental male was homozygous. If only some of the male progeny express the trait, the possibility still remains that the trait is Y-linked but incompletely penetrant. In either case, more conclusive evidence is needed.

Mate the female progeny (F_1 females) with normal males. If the trait is autosomal dominant, some of the male F_2 progeny will have enlarged testes, proving that the trait can be passed through a female. If the trait is Y-linked, all the male F_2 progeny will have normal testes, like their normal male father.

Chapter Five: Linkage, Recombination, and Eukaryotic Gene Mapping

COMPREHENSION QUESTIONS

Section 5.1

1. What does the term *recombination* mean? What are two causes of recombination?

   Solution:
   Recombination means that meiosis generates gametes with allelic combinations that differ from the original gametes inherited by the organism. Recombination may be caused by the independent assortment of loci on different chromosomes or by a physical crossing over between two loci on the same chromosome.

Section 5.2

2. What effect does crossing over have on linkage?

   Solution:
   Crossing over generates recombination between genes located on the same chromosome, and thus renders linkage incomplete.

3. Why is the frequency of recombinant gametes always half the frequency of crossing over?

   Solution:
   Crossing over occurs at the four-strand stage, when two homologous chromosomes, each consisting of a pair of sister chromatids, are paired. Each crossover involves just two of the four strands and generates two recombinant strands. The remaining two strands that were not involved in the crossover generate two nonrecombinant strands. Therefore, the frequency of recombinant gametes is always half the frequency of crossovers.

4. What is the difference between genes in coupling configuration and genes in repulsion? How does the arrangement of linked genes (whether they are in coupling or in repulsion) affect the results of a genetic cross?

   Solution:
   For genes in coupling configuration, two wild-type alleles are on the same chromosome and the two mutant alleles are on the homologous chromosome. For genes in repulsion, the wild-type allele of one gene and the mutant allele of the other gene are on the same chromosome and vice versa on the homologous chromosome. The two arrangements have opposite effects on the results of a cross. For genes in coupling configuration, most of the progeny will be either wild type for both genes or mutant for both genes, with relatively few that are wild type for one gene and mutant for the other. For genes in repulsion, most of the progeny will be mutant for only one gene and wild type for the other, with relatively few recombinants that are wild type for both or mutant for both.

5.   What is the difference between a genetic map and a physical map?

Solution:
A genetic map gives the order of genes and relative distance between them based on recombination frequencies observed in genetic crosses. A physical map locates genes on the actual chromosome or DNA sequence, and thus represents the physical distance between genes.

Section 5.3

6.   Explain how to determine, using the numbers of progeny from a three-point cross, which of three linked loci is the middle locus.

Solution:
Double crossovers always result in switching the middle gene with respect to the two nonrecombinant chromosomes. Hence, one can compare the two double crossover phenotypes with the two nonrecombinant phenotypes and see which gene is reversed. In the diagram below, we see that the coupling relationship of the middle gene is flipped in the double crossovers with respect to the genes on either side. Therefore, whichever gene on the double crossover can be altered to make the double crossover resemble a nonrecombinant chromosome is the middle gene. If we take either of the double crossover products *l M* r or *L m R*, changing the *M* gene will make it resemble a nonrecombinant.

7.   What does the interference tell us about the effect of one crossover on another?

Solution:
Positive interference indicates that a crossover inhibits or interferes with the occurrence of a second crossover nearby. Negative interference suggests that a crossover event can stimulate additional crossover events in the same region of the chromosome.

APPLICATION QUESTIONS AND PROBLEMS

Section 5.2

8.   In the snail *Cepaea nemoralis*, an autosomal allele causing a banded shell (B^B) is recessive to the allele for an unbanded shell (B^O). Genes at a different locus determine the background color of the shell; here, yellow (C^Y) is recessive to brown (C^{Bw}). A banded, yellow snail is crossed with a homozygous brown, unbanded snail. The F₁ are then crossed with banded, yellow snails (a testcross).

a. What will the results of the testcross be if the loci that control banding and color are linked with no crossing over?

Solution:
With absolute linkage, there will be no recombinant progeny. The F1 inherited banded and yellow alleles ($B^B C^Y$) together on one chromosome from the banded yellow parent and unbanded and brown alleles ($B^O C^{Bw}$) together on the homologous chromosome from the unbanded brown parent. Without recombination, all the F1 gametes will contain only these two allelic combinations, in equal proportions. Therefore, the F2 testcross progeny will be ½ banded, yellow and ½ unbanded, brown.

b. What will the results of the testcross be if the loci assort independently?

Solution:
With independent assortment, the progeny will be:
$\frac{1}{4}$ banded, yellow
$\frac{1}{4}$ banded, brown
$\frac{1}{4}$ unbanded, yellow
$\frac{1}{4}$ unbanded, brown

c. What will the results of the testcross be if the loci are linked and 20 m.u. apart?

Solution:
The recombination frequency is 20%, so each of the two classes of recombinant progeny must be 10%. The recombinants are banded, brown and unbanded, yellow. The two classes of nonrecombinants are 80% of the progeny, so each must be 40%. The nonrecombinants are banded, yellow and unbanded, brown.
In summary:
40% banded, yellow
40% unbanded, brown
10% banded, brown
10% unbanded, yellow

*9. In silkmoths (*Bombyx mori*), red eyes (*re*) and white-banded wings (*wb*) are encoded by two mutant alleles that are recessive to those that produce wild-type traits (*re*⁺ and *wb*⁺); these two genes are on the same chromosome. A moth homozygous for red eyes and white-banded wings is crossed with a moth homozygous for the wild-type traits. The F1 have normal eyes and normal wings. The F1 are crossed with moths that have red eyes and white-banded wings in a testcross. The progeny of this testcross are:

wild-type eyes, wild-type wings	418
red eyes, wild-type wings	19
wild-type eyes, white-banded wings	16
red eyes, white-banded wings	426

a. What phenotypic proportions would be expected if the genes for red eyes and for white-banded wings were located on different chromosomes?

Solution:
$\frac{1}{4}$ wild-type eyes, wild-type wings; $\frac{1}{4}$ red eyes, wild-type wings; $\frac{1}{4}$ wild-type eyes, white-banded wings; $\frac{1}{4}$ red eyes, white-banded wings.

b. What is the rate of recombination between the genes for red eyes and those for white-banded wings?

Solution:
The recombinant progeny are the 19 with red eyes, wild-type wings and 16 with wild-type eyes, white banded wings.

RF-recombinants/total progeny $\times$ 100% = (19 +16)/879 $\times$ 100% = 4.0%
The distance between the genes is 4 map units.

*10. A geneticist discovers a new mutation in *Drosophila melanogaster* that causes the flies to shake and quiver. She calls this mutation quiver (*qu*) and determines that it is due to an autosomal recessive gene. She wants to determine whether the gene encoding *quiver* is linked to the recessive gene for vestigial wings (*vg*). She crosses a fly homozygous for quiver and vestigial traits with a fly homozygous for the wild-type traits and then uses the resulting F₁ females in a testcross. She obtains the following flies from this testcross:

$vg^+ qu^+$	230
$vg\ qu$	224
$vg\ qu^+$	97
$vg^+ qu$	99
Total	650

Are the genes that cause vestigial wings and the quiver mutation linked? Do a chi-square test of independence to determine whether the genes have assorted independently.

Solution:
To test for independent assortment, we first test for equal segregation at each locus, then test whether the two loci sort independently.
Test for *vg*:
Observed vg = 224 + 97 = 321
Observed vg^+ = 230 + 99 = 329
Expected vg or vg^+ = ½ $\times$ 650 = 325

$$\chi^2 = \Sigma \frac{(\text{observed} - \text{expected})^2}{\text{expected}} = (321 - 325)^2/325 + (329 - 325)^2/325 = 16/325 + 16/325$$

$$= 0.098$$

We have $n-1$ degrees of freedom, where n is the number of phenotypic classes = 2, so just 1 degree of freedom. From **Table 3.4,** we see that the P value is between 0.7 and 0.8. So these results do not deviate significantly from the expected 1: 1 segregation.

Similarly, testing for qu, we observe 327 qu^+ and 323 qu and expect $\frac{1}{2} \times 650 = 325$ of each:

$\chi^2 = 4/325 + 4/325 = 0.025$, again with 1 degree of freedom. The P- value is between 0.8 and 0.9, so these results do not deviate significantly from the expected 1: 1 ratio.

Finally, we test for independent assortment, where we expect 1:1:1:1 phenotypic ratios, or 162.5 of each.

Observed	Expected	O – E	(O – E)²	(O – E)²/E
230	162.5	67.5	4556.25	28.0
224	162.5	61.5	3782.25	23.3
97	162.5	–65.5	4290.25	26.4
99	162.5	–63.5	4032.25	24.8

We have four phenotypic classes, giving us three degrees of freedom. The chi-square value of 102.5 is off the chart, so we reject independent assortment.

Instead, the genes are linked, and the $RF = (97 + 99)/650 \times 100\% = 30\%$, giving us 30 map units between them.

*11. In cucumbers, heart-shaped leaves (hl) are recessive to normal leaves (Hl) and having numerous fruit spines (ns) are recessive to few fruit spines (Ns). The genes for leaf shape and for number of spines are located on the same chromosome; findings from mapping experiments indicate that they are 32.6 m.u. apart. A cucumber plant having heart-shaped leaves and numerous spines is crossed with a plant that is homozygous for normal leaves and few spines. The F₁ are crossed with plants that have heart-shaped leaves and numerous spines. What phenotypes and phenotypic proportions are expected in the progeny of this cross?

Solution:
Because the genes for leaf shape and fruit spines are 32.6 m.u. apart, we expect that 32.5% of the progeny of the cross will be recombinants. There will be two types of recombinants (those with heart-shaped leaves and few spines and those with normal-shaped leaves and few spines), so each recombinant phenotype will constitute 16.3% of the progeny. The nonrecombinant progeny (those with heart-shaped leaves and numerous spines and those with normal-shaped leaves and few spines) will make up the remainder of the progeny (100% – 32.6% = 67.4%), which will be equally divided between the two nonrecombinant phenotypes (67.4/2 = 33.7% each).

Heart-shaped, numerous spines	33.7%
Normal-shaped, few spines	33.7%
Heart-shaped, few spines	16.3%
Normal-shaped, numerous spines	16.3%

12.   In tomatoes, tall (*D*) is dominant over dwarf (*d*), and smooth fruit (*P*) is dominant over pubescent (*p*) fruit, which is covered with fine hairs. A farmer has two tall and smooth tomato plants, which we will call plant A and plant B. The farmer crosses plants A and B with the same dwarf and pubescent plant and obtains the following numbers of progeny:

	Progeny of	
	Plant A	**Plant B**
Dd Pp	122	2
Dd pp	6	82
dd Pp	4	82
dd pp	124	4

a.   What are the genotypes of plant A and plant B?

Solution:
The genotypes of both plants are *Dd Pp*.

b.   Are the loci that determine the height of the plant and pubescence linked? If so, what is the percent recombination between them?

Solution:
Yes. From the cross of plant A, the map distance is 10/256 = 3.9%, or 3.9 m.u. The cross of plant B gives 6/170 = 3.5%, or 3.5 m.u. If we pool the data from the two crosses, we get 16/426 = 3.8%, or 3.8 m.u.

c.   Explain why different proportions of progeny are produced when plant A and plant B are crossed with the same dwarf pubescent plant.

Solution:
The two plants have different coupling configurations. In plant A, the dominant alleles *D* and *P* are coupled; one chromosome is *D P* and the other is *d p*. In plant B, they are in repulsion; its chromosomes have *D p* and *d P*.

13.   Alleles *A* and *a* are at a locus on the same chromosome as is a locus with alleles *B* and *b*. *Aa Bb* is crossed with *aa bb* and the following progeny are produced:

Aa Bb	5
Aa bb	45
aa Bb	45
aa bb	5

What conclusion can be made about the arrangement of the genes on the chromosome in the *Aa Bb* parent?

Solution:
The results of this testcross reveal that *Aa bb* and *aa Bb*, with far greater numbers, are the progeny that received nonrecombinant chromatids from the *Aa Bb* parent. Given that all the progeny received *ab* from the *aa bb* parent, the nonrecombinant progeny received

either an *Ab* or an *aB* chromatid from the *Aa Bb* parent. Therefore, the *A* and *B* loci are in repulsion in the *Aa Bb* parent. *Aa Bb* and *aa bb* are the recombinant classes, and their frequencies indicate that the genes *A* and *B* are 10 m.u. apart.

14.   Recombination frequencies between three loci in corn are shown here.

Loci	Recombination frequency (%)
R and W_2	17
R and L_2	35
W_2 and L_2	18

What is the order of the genes on the chromosome?

Solution:
The distances between the genes are indicated by the recombination rates. Because loci *R* and L_2 have the greatest recombination rate, they must be the furthest apart and W_2 is in the middle. The order of the genes is: *R*, W_2, L_2.

15.   In German cockroaches, bulging eyes (*bu*) are recessive to normal eyes (*bu$^+$*) and curved wings (*cv*) are recessive to straight wings (*cv$^+$*). Both traits are encoded by autosomal genes that are linked. A cockroach has genotype *bu$^+$ bu cv$^+$ cv* and the genes are in repulsion. Which of the following sets of genes will be found in the most common gametes produced by this cockroach?
a.   *bu$^+$ cv$^+$*
b.   *bu cv*
c.   *bu$^+$ bu*
d.   *cv$^+$ cv*
e.   *bu cv$^+$*
Explain your answer.

Solution:
The most common gametes will have (**e**) *bu cv$^+$*. Equally common will be gametes that have *bu$^+$ cv*, not given among the choices. Since these genes are linked, in repulsion, the wild-types alleles are on different chromosomes. Thus, the cockroach has one chromosome with *bu$^+$ cv* and the homologous chromosome with *bu cv$^+$*. Meiosis always produces nonrecombinant gametes at higher frequencies than recombinants, so gametes bearing *bu cv$^+$* will be produced at higher frequencies than (**a**) or (**b**), which are the products of recombination. The choices (**c**) and (**d**) have two copies of one locus and no copy of the other locus. They violate Mendelian segregation: Each gamete must contain one allele of each locus.

*16.   In *Drosophila melanogaster*, ebony body (*e*) and rough eyes (*ro*) are encoded by autosomal recessive genes found on chromosome 3; they are separated by 20 m.u. The gene that encodes forked bristles (*f*) is X-linked recessive and assorts independently of *e* and *ro*. Give the phenotypes of progeny and their expected proportions when a female of each of the following genotypes is test-crossed with a male.

a. $\dfrac{e^+ \qquad ro^+ \qquad f^+}{e \qquad ro \qquad f}$

Solution:
We can calculate the four phenotypic classes and their proportions for e and ro, and then each of those classes will be split 1:1 for f because f sorts independently. The recombination frequency between e and ro is 20%, so each of the recombinants (e^+ ro and e ro^+) will be 10%, and each of the nonrecombinants (e^+ ro^+ and e ro) will be 40%. Each of these will then be split equally among f^+ and f.

$e^+ ro^+ f^+$	20%
$e^+ ro^+ f$	20%
$e\ ro\ f^+$	20%
$e\ ro\ f$	20%
$e^+ ro\ f^+$	5%
$e^+ ro\ f$	5%
$e\ ro^+ f^+$	5%
$e\ ro^+ f$	5%

b. $\dfrac{e^+ \quad ro \quad f^+}{e \quad ro^+ \quad f}$

Solution:
We can do the same calculations as in part (**a**), except the nonrecombinants are e^+ ro and e ro^+ and the recombinants are e^+ ro^+ and e ro.

$e^+ ro^+ f^+$	5%
$e^+ ro^+ f$	5%
$e\ ro\ f^+$	5%
$e\ ro\ f$	5%
$e^+ ro\ f^+$	20%
$e^+ ro\ f$	20%
$e^+ ro^+ f^+$	20%
$e\ ro^+ f$	20%

17. Perform a chi-square test of independence on the data provided in **Figure 5.2** to determine if the genes for flower color and pollen shape in sweet peas are assorting independently. Give the chi-square value, degrees of freedom, and associated probability. What conclusion would you make about the independent assortment of these genes?

Solution:
One way is to test for a goodness-of-fit to a 9:3:3:1 ratio.

Phenotype	Observed	Expected	$(O-E)^2/E$
Purple, long	284	214.3	22.7
Purple, round	21	71.4	35.6
Red, long	21	71.4	35.6
Red, round	55	23.8	40.8

The chi-square value = 134.7, $df = 3$, $P < 0.001$.

Another way is to perform a chi-square test for independence of flower color and pollen shape, using a 2×2 contingency table:

Observed	Long	Round	Total
Purple	284	21	305
Red	21	55	76
Total	305	76	381

chi-square $= 381[(284)(55) - (21)(21)]^2/[(305)(76)(76)(305)]$
chi-square value $= 163.38$, $df = 1$, $P < 0.001$. It is extremely unlikely that the genes for flower color and pollen shape are assorting independently.

*18. A series of two-point crosses were carried out among seven loci (a, b, c, d, e, f, and g), producing the following recombination frequencies. Map the seven loci, showing their linkage groups, the order of the loci in each linkage group, and distances between the loci of each group.

Loci	Recombination frequency (%)	Loci	Recombination frequency (%)
a and b	50	c and d	50
a and c	50	c and e	26
a and d	12	c and f	50
a and e	50	c and g	50
a and f	50	d and e	50
a and g	4	d and f	50
b and c	10	d and g	8
b and d	50	e and f	50
b and e	18	e and g	50
b and f	50	f and g	50
b and g	50		

Solution:
50% recombination indicates that the genes assort independently. Less than 50% recombination indicates linkage. Starting with the most tightly linked genes a and g, we look for other genes linked to these and find only gene d has less than 50% recombination with a and g. So one linkage group consists of a, g, and d. We know that gene g is between a and d because the a to d distance is 12.

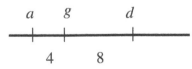

Similarly, we find a second linkage group of b, c, and e, with b in the middle.

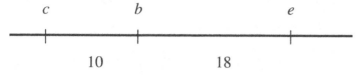

Gene f is unlinked to either of these groups; it is on a third linkage group.

19.   [Data Analysis Problem] R. W. Allard and W. M. Clement determined recombination rates for a series of genes in lima beans (R. W. Allard and W. M. Clement. 1959. *Journal of Heredity* 50:63–67). The following table lists paired recombination rates for eight of the loci (D, Wl, R, S, L_1, Ms, C, and G) that they mapped. On the basis of these data, draw a series of genetic maps for the different linkage groups of the genes, indicating the distances between the genes. Keep in mind that these frequencies are estimates of the true recombination frequencies and that some error is associated with each estimate. An asterisk beside a recombination frequency indicates that the recombination frequency is significantly different from 50%.

Recombination frequencies (%) among seven loci in lima beans

	Wl	*R*	*S*	*L₁*	*Ms*	*C*	*G*
D	2.1*	39.3*	52.4	48.1	53.1	51.4	49.8
Wl		38.0*	47.3	47.7	48.8	50.3	50.4
R			51.9	52.7	54.6	49.3	52.6
S				26.9*	54.9	52.0	48.0
L₁					48.2	45.3	50.4
Ms						14.7*	43.1
C							52.0

*Significantly different from 50%.

Solution:
Genes that show significantly less than 50% recombination frequency are linked. From this table we see that D, Wl, and R are linked, S and L_1 are linked, Ms and C are linked, and G appears to be unlinked to the other genes. Of the three linked loci, D and R appear to be the farthest apart, so that Wl is between D and R. We then use the recombination frequency as a measure of approximate distance in drawing the linkage maps.

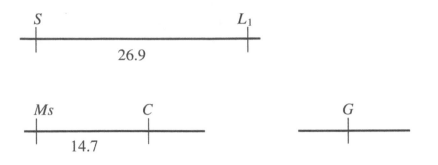

Section 5.3

*20. Waxy endosperm (*wx*), shrunken endosperm (*sh*), and yellow seedlings (*v*) are encoded by three recessive genes in corn that are linked on chromosome 5. A corn plant homozygous for all three recessive alleles is crossed with a plant homozygous for all the dominant alleles. The resulting F_1 are then crossed with a plant homozygous for the recessive alleles in a three-point testcross. The progeny of the testcross are:

wx	*sh*	*V*	87
Wx	*Sh*	*v*	94
Wx	*Sh*	*V*	3479
wx	*sh*	*v*	3478
Wx	*sh*	*V*	1515
wx	*Sh*	*v*	1531
wx	*Sh*	*V*	292
Wx	*sh*	*v*	280
Total			10,756

a. Determine order of these genes on the chromosome.

Solution:
The nonrecombinants are *Wx Sh V* and *wx sh v*.
The double crossovers are *wx sh V* and *Wx Sh v*.
Comparing the two, we see that they differ only at the *v* locus, so *v* must be the middle gene.

b. Calculate the map distances between the genes.

Solution:
Wx-V distance—recombinants are *wx V* and *Wx v*:
RF = (292 + 280 + 87 + 94)/10,756 = 753/10,756 = 0.07 = 7%, or 7 m.u.
Sh-V distance—recombinants *are sh V* and *Sh v*:
RF = (1515 + 1531 + 87 + 94)/10,756 = 3227/10,756 = 30 = 30%, or 30 m.u.
The *Wx–Sh* distance is the sum of these two distances: 7 + 30 = 37 m.u.l.

c. Determine the coefficient of coincidence and the interference among these genes.

Solution:
Expected dcos = $RF1 \times RF2 \times$ total progeny = 0.07(0.30)(10,756) = 226
C.o.C. = observed dcos/expected dcos = (87 + 94)/226 = 0.80
Interference = 1 − C.o.C = 0.20

21.   [Data Analysis Problem] Priscilla Lane and Margaret Green studied the linkage relations of three genes affecting coat color in mice: mahogany (*mg*), agouti (*a*), and ragged (*Rg*). They carried out a series of three-point crosses, mating mice that were heterozygous at all three loci with mice that were homozygous for the recessive alleles at these loci (P. W. Lane and M. C. Green. 1960. *Journal of Heredity* 51:228–230). The following table lists the progeny of the testcrosses.

Phenotype			Number
a	*Rg*	+	1
+	+	*mg*	1
a	+	+	15
+	*Rg*	*mg*	9
+	+	+	16
a	*Rg*	*mg*	36
a	+	*mg*	76
+	*Rg*	+	69
Total			213

Note: + represents a wild-type allele.

a. Determine the order of the loci that encode mahogany, agouti, and ragged on the chromosome, the map distances between them, and the interference and coefficient of coincidence for these genes.

Solution:
Based on the numbers of progeny, the nonrecombinants are *a* + *mg* and + *Rg* +. The double crossovers are *a Rg* + and + + *mg*. These differ in the *a* locus; therefore, *a* (agouti) is the middle locus.

The recombinants between *a* and *mg* are *a* and + or + and *mg* (ignoring the *Ra* locus): *RF* = (1 + 1 + 15 + 9)/213 = 0.12 = 12%, or 12 m.u.
The recombinants between *a* and *Rg* are + + or *a Rg* (ignoring the mg locus). *RF* = (1 + 1 + 16 + 36)/213 = 0.25 = 25%, or 25 m.u.

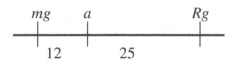

The C.o.C = observed dcos/expected dcos = 2/(0.12)(0.25)(213) = 2/6.4 = 0.31.
Interference = 1 − C.o.C. = 1 − 0.31 = 0.69

b. Draw a picture of the two chromosomes in the triply heterozygous mice used in the testcrosses, indicating which of the alleles are present on each chromosome.

Solution:
The parental chromosomes have the nonrecombinant coupling relationships: *a + mg* and *+ Rg +*. Shown with a as the middle locus:

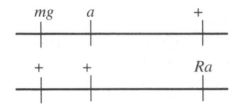

22.   Fine spines (*s*), smooth fruit (*tu*), and uniform fruit color (*u*) are three recessive traits in cucumbers, the genes for which are linked on the same chromosome. A cucumber plant heterozygous for all three traits is used in a testcross, and the following progeny are produced from this testcross:

S	U	Tu	2
s	u	Tu	70
S	u	Tu	21
s	u	tu	4
S	U	tu	82
s	U	tu	21
s	U	Tu	13
S	u	tu	17
Total			230

a. Determine the order of these genes on the chromosome.

Solution:
Nonrecombinants are *s u Tu* and *S U tu*.
Double crossovers are *s u tu* and *S U Tu*.
Because *Tu* differs between the nonrecombinants and the double crossovers, *Tu* is the middle gene.

b. Calculate the map distances between the genes.

Solution:
S–Tu distance: recombinants are *S Tu* and *s tu*.
RF = (2 + 4 + 21 + 21)/230 = 48/230 = 21%, or 21 m.u.
U–Tu distance: recombinants are *u tu* and *U Tu*.
RF = (2 + 4 + 13 + 17)/230 = 36/230 = 16%, or 16 m.u.

c. Determine the coefficient of coincidence and the interference among these genes.

Solution:
Expected dcos = (48/230)(36/230)(230) = 7.5
C.o.C. = observed dcos/expected dcos = 6/7.5 = 0.8
I = 1 – C.o.C. = 0.2

d. List the genes found on each chromosome in the parents used in the testcross.

Solution:
In the correct gene order for the heterozygous parent: *s Tu u* and *S tu U* For the testcross parent: *s tu u* and *s tu u*

*23. [Data Analysis Problem] Raymond Popp studied linkage among genes for pink eye (*p*), shaker-1 (*sh*-1, which causes circling behavior, head tossing, and deafness), and hemoglobin (*Hb*) in mice (R. A. Popp. 1962. *Journal of Heredity* 53:73–80). He performed a series of testcrosses, in which mice heterozygous for pink-eye, shaker-1, and hemoglobin 1 and 2 were crossed with mice that were homozygous for pink-eye, shaker-1, and hemoglobin 2.

$$\frac{P\ Sh\text{-}1\ Hb^1}{p\ sh\text{-}1\ Hb^2} \times \frac{p\ sh\text{-}1\ Hb^2}{p\ sh\text{-}1\ Hb^2}$$

The following progeny were produced:

Progeny genotype	Number
$\dfrac{p\ sh\text{-}1\ Hb^2}{p\ sh\text{-}1\ Hb^2}$	274
$\dfrac{P\ Sh\text{-}1\ Hb^1}{p\ sh\text{-}1\ Hb^2}$	320
$\dfrac{P\ sh\text{-}1\ Hb^2}{p\ sh\text{-}1\ Hb^2}$	57
$\dfrac{p\ Sh\text{-}1\ Hb^1}{p\ sh\text{-}1\ Hb^2}$	45
$\dfrac{P\ Sh\text{-}1\ Hb^2}{p\ sh\text{-}1\ Hb^2}$	6

$$\frac{P \; Sh\text{-}1 \; Hb^1}{p \; sh\text{-}1 \; Hb^2} \qquad\qquad 5$$

$$\frac{p \; Sh\text{-}1 \; Hb^2}{p \; sh\text{-}1 \; Hb^2} \qquad\qquad 0$$

$$\frac{P \; sh\text{-}1 \; Hb^1}{p \; sh\text{-}1 \; Hb^2} \qquad\qquad 1$$

Total   708

a. Determine the order of these genes on the chromosome.

Solution:
All the progeny receive $p \; sh\text{-}1 \; Hb^2$ from the male parent, shown as the lower chromosome. The upper chromosomes are the products of meiosis in the heterozygous parent, and are identified above. The nonrecombinants have $p \; sh\text{-}1 \; Hb^2$ or $P \; Sh\text{-}1 \; Hb^1$; the double crossovers have $p \; Sh\text{-}1 \; Hb^2$ or $P \; sh\text{-}1 \; Hb^1$. The two classes differ in the $Sh\text{-}1$ locus; therefore, $Sh\text{-}1$ is the middle locus.

b. Calculate the map distances between the genes.

Solution:
P and $Sh\text{-}1$: recombinants have $P \; sh\text{-}1$ or $p \; Sh\text{-}1$. $RF = (57 + 45 + 1)/708 = 0.145 = $
    14.5 m.u.
$Sh\text{-}1$ and Hb: recombinants have $Sh\text{-}1 \; Hb^2$ or $sh\text{-}1 \; Hb^1$. $RF = (6 + 5 + 1)/708 = 0.017 = $
    1.7 m.u.

c. Determine the coefficient of coincidence and the interference among these genes.

Solution:
Expected dcos $= RF1 \times RF2 \times$ total progeny $= 0.145(0.017)(708) = 1.7$
C.o.C. $=$ observed dcos/expected dcos $= 1/1.7 = 0.59$
Interference $= 1 - $ C.o.C $= 0.41$

*24. In *Drosophila melanogaster*, black body (b) is recessive to gray body (b^+), purple eyes (pr) are recessive to red eyes (pr^+), and vestigial wings (vg) are recessive to normal wings (vg^+). The loci encoding these traits are linked, with the following map distances:

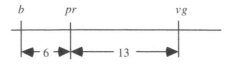

The interference among these genes is 0.5. A fly with a black body, purple eyes, and vestigial wings is crossed with a fly homozygous for a gray body, red eyes, and normal wings. The female progeny are then crossed with males that have a black body, purple eyes, and vestigial wings. If 1000 progeny are produced from this testcross, what will be the phenotypes and proportions of the progeny?

Solution:

Genotype	Body	Eyes	Wings	Proportion
$b^+ pr^+ vg^+$	normal	normal	normal	40.7%
$b\ pr\ vg$	black	purple	vestigial	40.7%
$b^+ pr^+ vg$	normal	normal	vestigial	6.3%
$b\ pr\ vg^+$	black	purple	normal	6.3%
$b^+ pr\ vg$	normal	purple	vestigial	2.8%
$b\ pr^+ vg^+$	black	normal	normal	2.8%
$b^+ pr\ vg^+$	normal	purple	normal	0.2%
$b\ pr^+ vg$	black	normal	vestigial	0.2%

25. *Sepia eyes, spineless bristles*, and *striped thorax* are three recessive mutations in *Drosophila* found on chromosome 3. A genetics student crosses a fly homozygous for sepia eyes, spineless bristles, and striped thorax with a fly homozygous for the wild-type traits—red eyes, normal bristles, and solid thorax. The female progeny are then test-crossed with males that have sepia eyes, spineless bristles, and striped thorax. Assume that the interference between these genes is 0.2 and that 400 progeny flies are produced from the testcross. Based on the map distances provided in **Figure 5.14**, predict the phenotypes and proportions of the progeny resulting from the testcross.

Solution:
According to **Figure 5.14**, *sepia* is at map position 26.0, *spineless* at 58.5, and *striped* at 62.0. The expected recombination frequencies should then be 32.5% between *sepia* and *spineless*, and 3.5% between *spineless* and *striped*.

First, the nonrecombinant phenotypes among the test-cross progeny will be the parental phenotypes: red eyes, normal bristles, solid thorax and sepia eyes, spineless bristles, and striped thorax. Because *spineless* is the middle gene, the double-crossover phenotypic classes among the test-cross progeny will be flies with red eyes, spineless bristles, and solid thorax, and flies with sepia eyes, normal bristles and striped thorax. Given the recombination frequencies and the interference value, we can calculate the number of double crossover progeny.
Interference = 0.2 = 1 – coefficient of coincidence
coefficient of coincidence = 0.8 = actual double crossovers/expected double crossovers

Actual double crossovers = 0.8 × expected double crossovers
Expected double crossovers = 0.325 × 0.035 × 400 = 4.55
Actual double crossovers = 0.8 (4.55) = 3.64; we round this up to four flies, split equally between the two double crossover classes.

The single crossovers between *sepia* and *spineless* would result in progeny with red eyes, spineless bristles, striped thorax, and the reciprocals with sepia eyes, normal bristles and solid thorax. To calculate the number of single crossovers, we remember that:

recombination frequency = *RF* = (single crossovers + double crossovers)/total progeny; single crossovers = *RF*(total progeny) – double crossovers = 0.325(400) – 4 = 126, split equally among the two phenotypic classes.
The single crossovers between *spineless* and *striped* would result in progeny with red eyes, normal bristles, striped thorax and the reciprocal sepia eyes, striped bristles and solid thorax. We calculate their numbers in the same way as above: single crossovers = *RF*(total progeny) – double crossovers = 0.035(400) – 4 = 10, split equally among the two classes.

The parental phenotypes are what are left of the 400 after subtracting the single crossovers and double crossovers from the 400 progeny.
parentals = 400 – (126 + 10 + 4) = 260, split equally among the two phenotypic classes.

In summary, the 400 progeny of the testcross will be split among eight phenotypic classes as follows:

Section 5.4

26.   Eight DNA sequences from different individuals are given in the diagram shown here.

		Nucleotide position			
	1	5	10	15	
Sequence 1	T C T G	G A T C	A T C A	C A T	. . .
Sequence 2	A C A G	C A T C	A T T A	C G T	. . .
Sequence 3	T C A G	G A T C	A T T A	C T A	. . .
Sequence 4	T C A G	G A T C	A T T A	C A T	. . .
Sequence 5	A C A G	C A T C	A T T A	C G T	. . .
Sequence 6	T C T G	G A T C	A T C A	C A T	. . .
Sequence 7	T C A G	G A T C	A T T A	C A T	. . .
Sequence 8	A C A G	C A T C	A T T A	C G T	. . .

a. Give the nucleotide positions of all single-nucleotide polymorphisms (SNPs; nucleotide positions where individuals vary in which base is present) in these sequences.

Solution:
SNPs occur at nucleotide positions 1, 3, 5, 11, 14, and 15.

b. How many different haplotypes (sets of linked variants) are found in these eight sequences?

Solution:
There are four haplotypes among the sequences.

c. Give the haplotype of each sequence by listing the specific bases at each variable position in that particular haplotype. See **Figure 20.8**.

Solution:
The haplotypes of each sequence are given below:
Sequence 1: T T G C A T   haplotype a
Sequence 2: A A C T G T   haplotype b
Sequence 3: T A G T T A   haplotype c
Sequence 4: T A G T A T   haplotype d
Sequence 5: A A C T G T   haplotype b
Sequence 6: T T G C A T   haplotype a
Sequence 7: T A G T A T   haplotype d
Sequence 8: A A C T G T   haplotype b

CHALLENGE QUESTION

Section 5.3

27. [Data Analysis Problem] Transferrin is a blood protein encoded by the transferrin locus (*Trf*). In house mice, the two alleles at this locus (*Trf*ᵃ and *Trf*ᵇ) are codominant and encode three types of transferrin:

Genotype	Phenotype
*Trf*ᵃ/*Trf*ᵃ	Trf-a
*Trf*ᵃ/*Trf*ᵇ	Trf-ab
*Trf*ᵇ/*Trf*ᵇ	Trf-b

The dilution locus, found on the same chromosome, determines whether the color of a mouse is diluted or full; an allele for dilution (*d*) is recessive to an allele for full color (d^+):

Genotype	Phenotype
$d^+ d^+$	d+ (full color)
$d^+ d$	d+ (full color)
dd	d (dilution)

Donald Shreffler conducted a series of crosses to determine the map distance between the transferrin locus and the dilution locus (D. C. Shreffler. 1963. *Journal of Heredity* 54:127–129). The following table presents a series of crosses carried out by Shreffler and the progeny resulting from these crosses.

| | | | Progeny phenotypes | | | | |
| | | | d^+ Trf-ab | d^+ Trf-b | d Trf-ab | d Trf-b | Total |
Cross	♂	♀					
1	$\dfrac{d^+\ Trf^a}{d\ Trf^b}\times\dfrac{d\ Trf^b}{d\ Trf^b}$		32	3	6	21	62
2	$\dfrac{d\ Trf^b}{d\ Trf^b}\times\dfrac{d^+\ Trf^a}{d\ Trf^b}$		16	0	2	20	38
3	$\dfrac{d\ Trf^a}{d\ Trf^b}\times\dfrac{d\ Trf^b}{d\ Trf^b}$		35	9	4	30	78
4	$\dfrac{d\ Trf^b}{d\ Trf^b}\times\dfrac{d^+\ Trf^a}{d\ Trf^b}$		21	3	2	19	45
5	$\dfrac{d^+\ Trf^b}{d\ Trf^a}\times\dfrac{d\ Trf^b}{d\ Trf^b}$		8	29	22	5	64
6	$\dfrac{d\ Trf^b}{d\ Trf^b}\times\dfrac{d^+\ Trf^b}{d\ Trf^a}$		4	14	11	0	29

a. Calculate the recombinant frequency between the *Trf* and the *d* loci by using the pooled data from all the crosses.

Solution:
We sum all the recombinant progeny and divide by the total progeny of all six crosses:
$RF = (3 + 6 + 0 + 2 + 9 + 4 + 3 + 2 + 8 + 5 + 4 + 0)/316 = 46/316 = 0.146 = 15\%$

b. Which crosses represent recombination in male gamete formation and which crosses represent recombination in female gamete formation?

Solution:
Crosses representing recombination in males are those where the female parent is homozygous and the male parent is heterozygous: crosses 1, 3, and 5.
Crosses 2, 4, and 6 represent female recombination because the female parent is heterozygous and the male parent is homozygous.

c. On the basis of your answer to part *b*, calculate the frequency of recombination among male parents and female parents separately.

Solution:
RF for males = $(3 + 6 + 9 + 4 + 8 + 5)/(62 + 78 + 64) = 35/204 = 0.17 = 17\%$
RF for females = $(0 + 2 + 3 + 2 + 4 + 0)/(38 + 45 + 29) = 11/112 = 0.098 = 9.8\%$

d. Are the rates of recombination in males and females the same? If not, what might produce the difference?

Solution:
The rate of recombination between these two genes appears to be higher in males than in females, although a statistical test should be performed to determine the significance of the apparent difference. Such gender differences in recombination rates could arise from differences in gametogenesis (meiosis is arrested for prolonged periods in mammalian oogenesis), imprinting of nearby loci affecting chromatin structure and crossing over in the region, or differences in synapsis of chromosomes in male and female meioses (Lynn et al. 2005. *Am. J. Hum. Genet.* 77:670–675.)

Chapter Six: Chromosome Variation

COMPREHENSION QUESTIONS

Section 6.1

1. List the different types of chromosome mutations and define each one.

   Solution:
   Chromosome rearrangements:
   > Deletion: loss of a part of a chromosome
   > Duplication: addition of an extra copy of a part of a chromosome
   > Inversion: a part of the chromosome is reversed in orientation
   > Translocation: a part of one chromosome becomes incorporated into a different (nonhomologous) chromosome

   Aneuploidy: loss or gain of one or more chromosomes, causing the chromosome number to deviate from $2n$ or the normal euploid complement

   Polyploidy: gain of entire sets of chromosomes, causing chromosome number changes from $2n$ to $3n$ (triploid), $4n$ (tetraploid), and so on

Section 6.2

2. Why do extra copies of genes sometimes cause drastic phenotypic effects?

   Solution:
   The expression of some genes is balanced with the expression of other genes; the ratios of their gene products, usually proteins, must be maintained within a narrow range for proper cell function. Extra copies of one of these genes cause that gene to be expressed at proportionately higher levels, thereby upsetting the balance of gene products.

3. Draw a pair of chromosomes as they would appear during synapsis in prophase I of meiosis in an individual heterozygous for a chromosome duplication.

   Solution:
   In the figure below, adapted from **Figure 6.5b**, the vertical dashed lines denote the locations of the genes labeled A–G. The lower chromosome has duplicated a segment containing genes C, D, and E.

   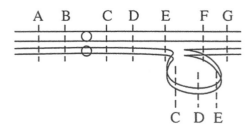

4. What is the difference between a paracentric and a pericentric inversion?

Solution:
A paracentric inversion does not include the centromere; a pericentric inversion includes the centromere.

5. How can inversions in which no genetic information is lost or gained cause phenotypic effects?

Solution:
Although inversions do not result in loss or duplication of chromosomal material, inversions can have phenotypic consequences if the inversion disrupts a gene at one of its breakpoints or if a gene near a breakpoint is altered in its expression because of a change in its chromosomal environment, such as relocation to a heterochromatic region. Such effects on gene expression are called position effects.

6. Explain why recombination is suppressed in individuals heterozygous for paracentric inversions.

Solution:
A crossover within a paracentric inversion produces a dicentric and an acentric recombinant chromatid. The acentric fragment is lost, and the dicentric fragment breaks, resulting in chromatids with large deletions that lead to nonviable gametes or embryonic lethality.

7. How do translocations in which no genetic information is lost or gained produce phenotypic effects?

Solution:
Translocations can produce phenotypic effects if the translocation breakpoint disrupts a gene or if a gene near the breakpoint is altered in its expression because of relocation to a different chromosomal environment (a position effect).

Section 6.3

8. List four major types of aneuploidy.

Solution:
Nullisomy: having no copies of a chromosome
Monosomy: having only one copy of a chromosome
Trisomy: having three copies of a chromosome
Tetrasomy: having four copies of a chromosome

9.   What is the difference between primary Down syndrome and familial Down syndrome?
      How does each type arise?

      Solution:
      Primary Down syndrome is caused by the spontaneous, random nondisjunction of
      chromosome 21, leading to trisomy 21. Familial Down syndrome most frequently arises as
      a result of a Robertsonian translocation of chromosome 21 with another chromosome,
      usually chromosome 14.

Section 6.4

10.   What is the difference between autopolyploidy and allopolyploidy? How does each arise?

      Solution:
      In autopolyploidy, all sets of chromosomes are from the same species. In allopolyploidy,
      the sets of chromosomes are derived from two or more different species. Autopolyploidy
      may arise through nondisjunction in an early $2n$ embryo or through meiotic nondisjunction
      that produces a gamete with extra sets of chromosomes. Allopolyploidy is usually
      preceded by hybridization between two different species, followed by chromosome
      doubling.

11.   Explain why autopolyploids are usually sterile, whereas allopolyploids are often fertile.

      Solution:
      Autopolyploids arise from duplication of their own chromosomes. During meiosis, the
      presence of more than two homologous chromosomes results in faulty alignment of
      homologues in prophase I, and subsequent faulty segregation of the homologues in
      anaphase I. The resulting gametes have an uneven distribution of chromosomes and are
      genetically unbalanced. These gametes usually produce lethal chromosome imbalances in
      the zygote. Allopolyploids, however, have chromosomes from different species. As long
      as they have a diploid set of chromosomes from each species, as in an allotetraploid or
      even an allohexaploid, the homologous chromosome pairs from each species can align and
      segregate properly during meiosis. Their gametes will be balanced and will produce viable
      zygotes when fused with other gametes from the same type of allopolyploid individual.

APPLICATION QUESTIONS AND PROBLEMS

Section 6.1

12.   Examine the karyotypes shown in **Figures 6.1** and **6.2a**. Are the individuals from whom
      these karyotypes were made males or females?

      Solution:
      Figure 6.1—male, because the karyotype has a Y chromosome. **Figure 6.2a**—female
      because the karyotype has two X chromosomes and no Y chromosome

*13. Which types of chromosome mutations

   a. increase the amount of genetic material in a particular chromosome?

   Solution:
   Duplications

   b. increase the amount of genetic material in all chromosomes?

   Solution:
   Polyploidy

   c. decrease the amount of genetic material in a particular chromosome?

   Solution:
   Deletions

   d. change the position of DNA sequences in a single chromosome without changing the amount of genetic material?

   Solution:
   Inversions

   e. move DNA from one chromosome to a nonhomologous chromosome?

   Solution:
   Translocations

Section 6.2

*14. A chromosome has the following segments, where • represents the centromere:
   AB•CDEFG

   What types of chromosome mutations are required to change this chromosome into each of the following chromosomes? (In some cases, more than one chromosome mutation may be required.)

   Solution:
   a. ABAB•CDEFG: Tandem duplication of AB
   b. AB•CDEABFG: Displaced duplication of AB
   c. AB•CFEDG: Paracentric inversion of DEF
   d. A•CDEFG: Deletion of B
   e. AB•CDE: Deletion of FG
   f. AB•EDCFG: Paracentric inversion of CDE
   g. C•BADEFG: Pericentric inversion of ABC
   h. AB•CFEDFEDG: Duplication and inversion of DEF
   i. AB•CDEFCDFEG: Duplication of CDEF, inversion of EF

15.   A chromosome initially has the following segments:
AB•CDEFG
Draw the chromosome, identifying its segments, that would result from each of the following mutations.

Solution:
a. Tandem duplication of DEF: AB•CDEFDEFG
b. Displaced duplication of DEF: AB•CDEFGDEF or other arrangements where the duplicated DEF is not adjacent to the original DEF
c. Deletion of FG: AB•CDE
d. Paracentric inversion that includes DEFG: AB•CGFED
e. Pericentric inversion of BCDE: AEDC•BFG

16.   The following diagram represents two nonhomologous chromosomes:
AB•CDEFG
RS•TUVWX
What type of chromosome mutation would produce each of the following chromosomes?

Solution:
a. AB•CD
RS•TUVWXEFG
Nonreciprocal translocation of EFG
b. AUVB•CDEFG
RS•TWX
Nonreciprocal translocation of UV
c. AB•TUVFG
RS•CDEWX
Reciprocal translocation of CDE and TUV
d. AB•CWG
RS•TUVDEFX
Reciprocal translocation of DEF and W

17.   The green-nose fly normally has six chromosomes: two metacentric and four acrocentric. A geneticist examines the chromosomes of an odd-looking green-nose fly and discovers that it has only five chromosomes; three of them are metacentric and two are acrocentric. Explain how this change in chromosome number might have taken place.

Solution:
A Robertsonian translocation between two of the acrocentric chromosomes would result in a new metacentric chromosome and a very small chromosome that may have been lost.

*18. A wild-type chromosome has the following segments:
ABC•DEFGHI

Researchers have found individuals that are heterozygous for each of the following chromosome mutations. For each mutation, sketch how the wild-type and mutated chromosomes would pair in prophase I of meiosis, showing all chromosome strands.

Solution:
a. ABC•DEFDEFGHI

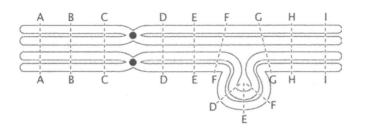

b. ABC•DHI

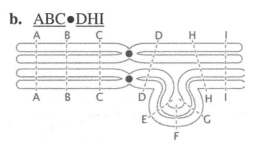

c. ABC•DGFEHI

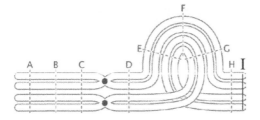

d. ABED•CFGHI

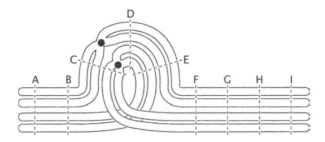

*19. [Data Analysis Problem] As discussed in this chapter, crossing over within a pericentric inversion produces chromosomes that have extra copies of some genes and no copies of other genes. The fertilization of gametes containing such duplication or deficient chromosomes often results in children with syndromes characterized by developmental delay, intellectual disability, the abnormal development of organ systems, and early death. Maarit Jaarola and colleagues examined individual sperm cells of a male who was heterozygous for a pericentric inversion on chromosome 8 and determined that crossing over took place within the pericentric inversion in 26% of the meiotic divisions (M. Jaarola, R. H. Martin, and T. Ashley. 1998. *American Journal of Human Genetics* 63:218–224).

Assume that you are a genetic counselor and that a couple seeks genetic counseling from you. Both the man and the woman are phenotypically normal, but the woman is heterozygous for a pericentric inversion on chromosome 8. The man is karyotypically normal. What is the probability that this couple will produce a child with a debilitating syndrome as the result of crossing over within the pericentric inversion?

Solution:
Each crossover event results in two recombinant and two nonrecombinant gametes. If one crossover occurs in 100% of meioses, the result would be 50% recombinant gametes. If crossing over occurs within the pericentric inversion at a rate of 26% of meioses, then 13% of the woman's oocytes will have duplication/deficient chromosome 8. If all of these oocytes form viable eggs, and if they do not result in early miscarriage after fertilization, then the probability of the couple having a child with a syndrome caused by the crossing over is 13%.

20. An individual heterozygous for a reciprocal translocation possesses the following chromosomes:

<u>AB•CDEFG</u>
<u>AB•CDVWX</u>
<u>RS•TUEFG</u>
<u>RS•TUVWX</u>

Draw the pairing arrangement of these chromosomes in prophase I of meiosis.

Solution:

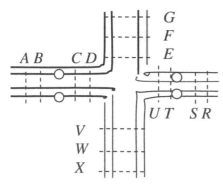

Section 6.3

*21. Red–green color blindness is a human X-linked recessive disorder. A young man with a 47,XXY karyotype (Klinefelter syndrome) is color blind. His 46,XY brother also is color blind. Both parents have normal color vision. Where did the nondisjunction that gave rise to the young man with Klinefelter syndrome take place? Assume that no crossing over took place in prophase I of meiosis.

Solution:
Because the father has normal color vision, the mother must be the carrier for color blindness. The color-blind young man with Klinefelter syndrome must have inherited two copies of the color-blind X chromosome from his mother. If no crossover occurred between the color blindness allele and the centromere, the nondisjunction event took place in meiosis II of the egg.

*22. Bill and Betty have had two children with Down syndrome. Bill's brother has Down syndrome and his sister has two children with Down syndrome. On the basis of these observations, indicate which of the following statements are most likely correct and which are most likely incorrect. Explain your reasoning.

   a. Bill has 47 chromosomes.
   b. Betty has 47 chromosomes.
   c. Bill and Betty's children each have 47 chromosomes.
   d. Bill's sister has 45 chromosomes.
   e. Bill has 46 chromosomes.
   f. Betty has 45 chromosomes.
   g. Bill's brother has 45 chromosomes.

Solution:
The high incidence of Down syndrome in Bill's family and among Bill's relatives is consistent with familial Down syndrome, caused by a Robertsonian translocation of chromosome 21. Bill and his sister, who are unaffected, are phenotypically normal carriers of the translocation and have 45 chromosomes. Their children and Bill's brother, who have Down syndrome, have 46 chromosomes. From the information given, there is no reason to suspect that Bill's wife Betty has any chromosomal abnormalities. Therefore, the statement in part *d* is most likely correct.

*23. In mammals, sex-chromosome aneuploids are more common than autosomal aneuploids, but in fishes, sex-chromosome aneuploids and autosomal aneuploids are found with equal frequency. Offer a possible explanation for these differences in mammals and fishes. (Hint: Think about why sex chromosome aneuploids are more common than autosomal aneuploids in mammals.)

Solution:
In mammals, the higher frequency of sex-chromosome aneuploids compared with that of autosomal aneuploids is due to X-chromosome inactivation and the lack of essential genes on the Y chromosome. If fish do not have X-chromosome inactivation and both of their sex chromosomes have numerous essential genes, then the frequency of aneuploids should be similar for both sex chromosomes and autosomes.

24.   [Data Analysis Problem] Using breeding techniques, Andrei Dyban and V.S. Baranov (*Cytogenetics of Mammalian Embryonic Development.* Oxford: Oxford University Press, Clarendon Press; New York: Oxford University Press, 1987) created mice that were trisomic for each of the different mouse chromosomes. They found that only mice with trisomy 19 completed development. Mice that were trisomic for all other chromosomes died in the course of development. For some of these trisomics, the researchers plotted the length of development (number of days after conception before the embryo died) as a function of the size of the mouse chromosome that was present in three copies (see the adjoining graph). Summarize their findings and provide a possible explanation for the results.

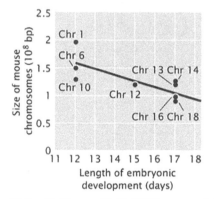

[After E. Torres, B. R. Williams, and A. Amon, *Genetics* 179:737–746, 2008.]

Solution:
Among the trisomics, they observed an inverse relationship between size of the chromosome present in three copies and the length of embryonic development. In other words, mice that were trisomic for small chromosomes developed longer than mice that were trisomic for large chromosomes. This can be explained by the fact that possessing extra chromosomes leads to abnormal gene dosage, which interferes with development. Larger chromosomes would most likely contain more genes. Being trisomic for larger chromosomes therefore affects the dosage of more genes and is more detrimental, causing development to cease at an earlier stage.

Section 6.4

25.   Species I has $2n = 16$ chromosomes. How many chromosomes will be found per cell in each of the following mutants in this species?

Solution:
a. Monosomic: 15
b. Autotriploid: 24
c. Autotetraploid: 32
d. Trisomic: 17
e. Double monosomic: 14
f. Nullisomic: 14
g. Autopentaploid: 40
h. Tetrasomic: 18

26.   Species I is diploid ($2n = 8$) with chromosomes AABBCCDD; related species II is diploid ($2n = 8$) with chromosomes MMNNOOPP. What types of chromosome mutations do the individuals with the following sets of chromosomes have?

Solution:
a. AAABBCCDD: Trisomy A
b. MMNNOOOOPP: Tetrasomy O
c. AABBCDD: Monosomy C
d. AAABBBCCCDDD: Triploidy
e. AAABBCCDDD: Ditrisomy A and D
f. AABBDD: Nullisomy C
g. AABBCCDDMMNNOOPP: Allotetraploidy
h. AABBCCDDMNOP: Allotriploidy

*27.   Species I has $2n = 8$ chromosomes and species II has $2n = 14$ chromosomes. What would the expected chromosome numbers be in individuals with the following chromosome mutations? Give all possible answers.

Solution:
a. Allotriploidy including species I and II
Such allotriploids could have $1n$ from species I and $2n$ from species II for $3n = 18$; alternatively, they could have $2n$ from species I and $1n$ from species II for $3n = 15$.
b. Autotetraploidy in species II
$4n = 28$
c. Trisomy in species I
$2n + 1 = 9$
d. Monosomy in species II
$2n - 1 = 13$
e. Tetrasomy of species I
$2n + 2 = 10$
f. Allotetraploidy including species I and II
$2n + 2n = 22$; $1n + 3n = 25$; $3n + 1n = 19$

28.   Suppose that species I in **Figure 6.25** has $2n = 10$ and species II in the figure had $2n = 12$. How many chromosomes would be present in the allotetraploid at the bottom of the figure?

Solution:
22

29.   Consider a diploid cell that has $2n = 4$ chromosomes: one pair of metacentric chromosomes and one pair of acrocentric chromosomes. Suppose that this cell undergoes nondisjunction, giving rise to an autotriploid cell ($3n$). The triploid cell then undergoes meiosis. Draw the different types of gametes that could result from meiosis in the triploid cell, showing the chromosomes present in each type. To distinguish between the different metacentric and acrocentric chromosomes, use a different color to draw each metacentric chromosome; similarly, use a different color to draw each acrocentric chromosome. (*Hint:* See **Figure 6.24**.)

Solution:

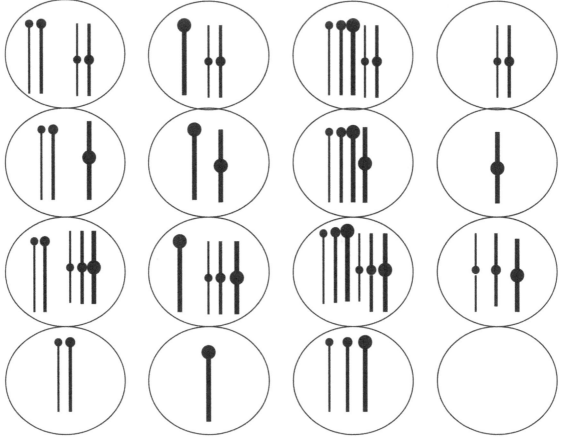

For each chromosome, the types of gametes that can result will have 1, 2, 3, or 0. Given two chromosomes, there are 16 basic combinations of the two chromosomes in the gametes. If the homologues are genetically distinct (as represented by different colors in the figure), there are many more possible genetic combinations of each chromosome (different colors combinations). Note that these combinations will not arise with equal frequencies; most trivalents will segregate 2:1, so combinations that result from 3:0 segregation will be less frequent.

30.   Assume that the autotriploid cell in **Figure 6.24** has $3n = 30$ chromosomes. For each of the gametes produced by this cell, give the chromosome number of the zygote that would result if the gamete fused with a normal haploid gamete.

Solution:
30, 20, 20, 30, 40, 10

31.   [Data Analysis Problem] *Nicotiana glutinosa* ($2n = 24$) and *N. tabacum* ($2n = 48$) are two closely related plants that can be intercrossed, but the F_1 hybrid plants that result are usually sterile. In 1925, Roy Clausen and Thomas Goodspeed crossed *N. glutinosa* and *N. tabacum,* and obtained one fertile F_1 plant (R. E. Clausen and T. H. Goodspeed. 1925. *Genetics* 10:278–284). They were able to self-pollinate the flowers of this plant to produce an F_2 generation. Surprisingly, the F_2 plants were fully fertile and produced viable seeds. When Clausen and Goodspeed examined the chromosomes of the F_2 plants, they observed 36 pairs of chromosomes in metaphase I and 36 individual chromosomes in metaphase II. Explain the origin of the F_2 plants obtained by Clausen and Goodspeed and the numbers of chromosomes observed.

Solution:
The rare fertile F_1 plant is most likely an allotetraploid. The initial fusion of haploid gametes forms a hybrid containing 12 *glutinosa* chromosomes and 24 *tabacum* chromosomes. A mitotic nondisjunction results in an allotetraploid containing 12 pairs of *glutinosa* chromosomes and 24 pairs of *tabacum* chromosomes. In the allotetraploid, meiosis proceeds normally, since each chromosome can pair with a homologue.

32.   What would be the chromosome number of progeny resulting from the following crosses in wheat (see **Figure 6.26**)? What type of polyploid (allotriploid, allotetraploid, etc.) would result from each cross?

Solution:
a. Einkorn wheat and emmer wheat
Einkorn is *T. monococcum* ($2n = 14$); emmer is *T. turgidum* ($4n = 28$). Gametes from einkorn have $n = 7$ and gametes from emmer have $2n = 14$ chromosomes. The progeny from this cross would be $3n = 21$ and allotriploid.
b. Bread wheat and emmer wheat
Bread wheat (*T. aestivum*) is allohexaploid ($6n = 42$) and produces gametes with $3n = 21$ chromosomes. Progeny from this cross would have ($3n = 21$) + ($2n = 14$), or $5n = 35$ chromosomes, and would be allopentaploid.
c. Einkorn wheat and bread wheat
($n = 7$) + ($3n = 21$) produces progeny with $4n = 28$ chromosomes, and are allotetraploid.

33.   [Data Analysis Problem] Karl and Hally Sax crossed *Aegilops cylindrica* ($2n = 28$), a wild grass found in the Mediterranean region, with *Triticum vulgare* ($2n = 42$), a type of wheat (K. Sax and H. J. Sax. 1924. *Genetics* 9:454–464). The resulting F_1 plants from this cross had 35 chromosomes. Examination of metaphase I in the F_1 plants revealed the presence of 7 pairs of chromosomes (bivalents) and 21 unpaired chromosomes (univalents).

    a. If the unpaired chromosomes segregate randomly, what possible chromosome numbers will appear in the gametes of the F_1 plants?

    Solution:
    Anywhere from 7 to 28. Each gamete will get 7 chromosomes from segregation of the bivalents, plus 0 to 21 univalents.

    b. What does the appearance of the bivalents in the F_1 hybrids suggest about the origin of *Triticum vulgare* wheat?

    Solution:
    The 7 bivalents suggest that *Triticum vulgare* has a common ancestor with *Aegilops* and that *Triticum* obtained 14 additional chromosomes from other ancestors not shared with *Aegilops*.

CHALLENGE QUESTIONS

Section 6.3

34.   Red–green color blindness is a human X-linked recessive disorder. Jill has normal color vision, but her father is color blind. Jill marries Tom, who also has normal color vision. Jill and Tom have a daughter who has Turner syndrome and is color blind.

    a. How did the daughter inherit color blindness?

    Solution:
    The daughter, with Turner syndrome, is 45,XO. A normal egg cell with a color-blind X chromosome was fertilized by a sperm carrying no sex chromosome. Such a sperm could have been produced by a nondisjunction event during spermatogenesis.

    b. Did the daughter inherit her X chromosome from Jill or from Tom?

    Solution:
    The color-blind daughter must have inherited her X chromosome from Jill. Tom is not color blind and therefore could not have a color-blind allele on his single X chromosome. Jill's father was color blind, so Jill must have inherited a color-blind X chromosome from him and passed it on to her daughter.

35. [Data Analysis Problem] Mules result from a cross between a horse ($2n = 64$) and a donkey ($2n = 62$), have 63 chromosomes, and are almost always sterile. However, in the summer of 1985, a female mule named Krause who was pastured with a male donkey gave birth to a foal (O. A. Ryder et al. 1985. *Journal of Heredity* 76:379–381). Blood tests established that the male foal, appropriately named Blue Moon, was the offspring of Krause and that Krause was indeed a mule. Both Blue Moon and Krause were fathered by the same donkey. The foal, like his mother, had 63 chromosomes—half of them horse chromosomes and the other half donkey chromosomes. Analyses of genetic markers showed that, remarkably, Blue Moon seemed to have inherited a complete set of horse chromosomes from his mother, instead of the random mixture of horse and donkey chromosomes that would be expected with normal meiosis. Thus, Blue Moon and Krause were not only mother and son, but also brother and sister.

a. With the use of a diagram, show how, if Blue Moon inherited only horse chromosomes from his mother, Krause and Blue Moon are mother and son as well as brother and sister.

Solution:

H1 represents the set of maternal horse chromosomes from Krause's mother; D1a represents the set of paternal donkey chromosomes from Krause's father. D1b represents a different set of paternal donkey chromosomes from Krause's (and Blue Moon's) father.

Krause and Blue Moon have the same father, and the same genetic mother, so they are siblings, although Krause is the mother of Blue Moon. In fact, they are closer than full siblings because Krause and Blue Moon have the exact same maternal horse chromosomes—they are half-clones, differing only in which set of chromosomes they received from the common father.

b. Although rare, additional cases of fertile mules giving births to offspring have been reported. In these cases, when a female mule mates with a male horse, the offspring is horselike in appearance, but when a female mule mates with a male donkey, the offspring is mulelike in appearance. Is this observation consistent with the idea that the offspring of fertile female mules inherit only a set of horse chromosomes from their mule mothers? Explain your reasoning.

Solution:
Yes. Mules result when a donkey sperm fertilizes a horse egg. The reverse, a horse sperm fertilizing a donkey egg, produces a hinny, which is similar in appearance to a mule. The production of horselike progeny when a female mule is mated with a male horse suggests that the mule's eggs largely lack donkey chromosomes. The observation that mating of the fertile female mule with a male donkey results in mulelike foals again suggests that the maternal chromosomes must be largely of horse origin.

c. Can you suggest a possible mechanism for how fertile female mules might pass on a complete set of horse chromosomes to their offspring?

Solution:
During oogenesis, meiotic divisions are unequal, resulting in a large egg cell and three small polar bodies. The key event is the first meiotic division. Somehow, a complete haploid set of horse chromosomes all segregate to the oocyte and not into the polar body. One possible mechanism is that horse chromosomes undergo an extra round of premeiotic replication, so the premeiotic cell has $2n$ horse chromosomes and $1n$ donkey chromosomes. Then the horse chromosomes pair and segregate, resulting in a complete set of horse chromosomes in the oocyte. Another possibility is that horse and donkey chromosomes pair, and the meiotic spindle apparatus interacts differentially with horse and donkey chromosomes so that horse chromosomes preferentially segregate to the oocyte. A third possibility is that horse and donkey chromosomes do not synapse with each other, and horse chromosomes end up in the oocyte because the oocyte volume is so much larger. Donkey chromosomes are lost in the polar body because they have a more peripheral localization than horse chromosomes.

Section 6.4

36.   Humans and many other complex organisms are diploid, possessing two sets of genes, one inherited from the mother and one from the father. However, a number of eukaryotic organisms spend most of their life cycles in a haploid state. Many of these eukaryotes, such as *Neurospora* and yeast, still undergo meiosis and sexual reproduction, but most of the cells that make up the organism are haploid.

Considering that haploid organisms are fully capable of sexual reproduction and generating genetic variation, why are most complex eukaryotes diploid? In other words, what might be the evolutionary advantage of existing in a diploid state instead of a haploid state? And why might a few organisms, such as *Neurospora* and yeast, exist as haploids?

Solution:
The most obvious advantage to being a diploid is genetic redundancy. Most mutations reduce or eliminate gene function, and are recessive. Having two copies means that the

organism, and its component cells, are able to survive the vast majority of mutations that would be lethal or deleterious in a haploid state. This redundancy is especially important for organisms that have large, complex genomes, complex development, and relatively lengthy life cycles.

Another advantage to diploidy is that gene expression levels can be higher than in haploid cells, often leading to larger cell sizes and larger, more robust organisms. Indeed, many cultivated polyploid crop plants have higher growth rates and yields than their diploid relatives. A third advantage is that the ability to carry recessive mutations in a masked state allows diploid populations to accumulate and harbor much more genetic diversity. Variant forms of genes that may be harmful to the organism and selected against in the current environment may prove advantageous when the environment changes.

These advantages are less important for organisms that have small genomes and simpler, shorter life cycles. Yeast and *Neurospora* genomes are highly adapted to their ecological niches, and haploid growth exposes and weeds out less favorable genetic variants. The ability to replicate their haploid genomes more quickly may give them a selective advantage over diploids. Indeed, these species become diploid and undergo meiosis only when conditions become less favorable for growth.

Chapter Seven: Bacterial and Viral Genetic Systems

COMPREHENSION QUESTIONS

Section 7.1

1. What is the difference between complete medium and minimal medium? How are complete media and minimal media to which one or more nutrients have been added (selective media) used to isolate auxotrophic mutants of bacteria?

    Solution:
    Auxotrophic mutants lack one or more enzymes necessary for synthesizing some essential molecule, such as an amino acid. Complete medium contains all the substances required by bacteria for growth and reproduction. Minimal medium contains only nutrients necessary for prototrophic (wild-type) bacteria. To isolate auxotrophic mutants that require leucine, bacteria are first grown on complete medium. Bacterial colonies that grow are transferred to plates with minimal medium to which leucine has been added and plates with minimal medium that lacks leucine. Any colony that grows on the plate containing leucine but fails to grow on the plate without leucine contains a leu-mutant that cannot synthesize leucine.

Section 7.2

2. Briefly explain the differences between F^+, F^-, Hfr, and F' cells.

    Solution:
    An F^+ cell will contain the F factor as a circular plasmid separate from the chromosome. The Hfr cell has the F factor integrated into its chromosome. In F' strains, the F factor exists as a separate circular plasmid, but the plasmid carries bacterial genes that were originally part of the bacterial chromosome. The F^- strain does not contain the F factor and can receive DNA from cells that contain the F factor (F^+, Hfr, and F' cells).

3. What types of matings are possible between F^+, F^-, Hfr, and F' cells? What outcomes do these matings produce? What is the role of the F factor in conjugation?

    Solution:

Types of matings	Outcomes
$F^+ \times F^-$	Two F^+ cells
Hfr $\times F^-$	One F^+ cell and one F^- cell
$F' \times F^-$	Two F' cells

    The F factor contains a number of genes involved in the conjugation process, including genes necessary for the synthesis of the sex pilus. The F factor has an origin of replication that enables the factor to be replicated in the conjugation process and genes for opening the plasmid and initiating the chromosome transfer.

4.   Explain how interrupted conjugation, transformation, and transduction can be used to map bacterial genes. How are these methods similar and how are they different?

Solution:
To map genes by conjugation, an Hfr strain is mixed with an F⁻ strain. The conjugation process is interrupted at regular intervals. The chromosomal transfer from the Hfr strain always begins with a part of the integrated F factor and proceeds in a linear fashion. The time required for individual genes to be transferred is relative to their position on the chromosome. Gene distances are typically mapped in minutes of conjugation.

In transformation, the recipient cell takes up the DNA directly from the environment. The relative frequency at which pairs of genes are transferred, or cotransformed, indicates the distance between the two genes. Closer gene pairs are cotransformed more frequently.

The transfer of DNA by transduction requires a viral vector. DNA from the donor cell is packaged into a viral protein coat. The viral particle containing the bacterial donor DNA then infects the recipient bacterial cell. The donor bacterial DNA is incorporated into the recipient cell's chromosome by recombination. Only genes that are close together on the bacterial chromosome can be cotransduced. Therefore, the rate of cotransduction, like the rate of cotransformation, is an indication of the physical distances between genes on the chromosome.

All three processes require the uptake by the recipient cell of a piece of the donor chromosome and the incorporation of some of that piece into the recipient chromosome by recombination. In all three processes, the mapping distance is calculated by measuring the frequency with which recipient cells are genetically altered. The processes use different methods to transfer donor DNA into the recipient cell.

Section 7.3

5.   List some of the characteristics that make bacteria and viruses ideal organisms for many types of genetic studies.

Solution:
(1) Reproduction is rapid, asexual, and produces lots of progeny.
(2) Their genomes are small.
(3) They are easy to grow in the laboratory.
(4) Techniques are available for isolating and manipulating their genes.
(5) Mutant phenotypes, especially auxotrophic phenotypes, are easy to measure.

6.   What types of genomes do viruses have?

Solution:
Viral genomes can consist of either DNA or RNA molecules. The viral nucleic acids can be either double stranded or single stranded, depending on the type of virus.

7.   Briefly describe the differences between the lytic cycle of virulent phages and the lysogenic cycle of temperate phages.

Solution:
Virulent phages reproduce strictly by the lytic cycle and ultimately result in the death of the host bacterial cell. During the lytic cycle, a virus injects its genome into the host cell. The genome directs production and assembly of new viral particles. A viral enzyme is produced and breaks open the cell, releasing new viral particles into the environment.

Temperate phages can utilize either the lytic or lysogenic cycle. The infection cycle begins when a viral particle injects its genome into the host cell. In the lysogenic cycle, the viral genome integrates into the host chromosome as a prophage. The inactive prophage can remain part of the bacterial chromosome for an extended period and is replicated along with the bacterial chromosome prior to cell division. Certain environmental stimuli can trigger the prophage to exit the lysogenic cycle and enter the lytic cycle.

8.   Briefly explain how genes in phages are mapped.

Solution:
To map genes in phages, bacterial cells are doubly infected with phage particles that differ in two or more genes. During the production of new phage progeny, the phage DNAs can undergo recombination, thus resulting in the formation of recombinant plaques. The rate of recombination is used to determine the linear order and relative distances between genes. The farther apart two genes are on the chromosome, the more frequently they will recombine.

9.   Briefly describe the genetic structure of a typical retrovirus.

Solution:
Retroviral genomes all have three genes in common: *gag, pol, and env.* Proteins that make up the viral capsid are encoded by the *gag* gene. Reverse transcriptase and an enzyme called integrase are encoded by the *pol* gene. While reverse transcriptase synthesizes double stranded viral DNA from an RNA template, integrase results in the insertion of the viral DNA into the host chromosome. Finally, the *env* gene encodes for proteins found on the viral envelope.

10.   What are the evolutionary origins of HIV-1 and HIV-2?

Solution:
Both HIV-1 and HIV-2 are related to simian immunodeficiency viruses (SIV). Analysis of DNA sequences indicates that HIV-1 is related to simian immunodeficiency virus found in chimpanzees (SIV_{cpz}). DNA sequence analysis also reveals that SIV_{cpz} is a hybrid virus created by recombination between a retrovirus in red-capped mangabey and a retrovirus found in the great spot monkey. HIV-2 sequence analysis shows that it evolved from a simian immunodeficiency virus found in sooty mangabeys (SIV_{sm}).

11.   Most humans are not easily infected by avian influenza. How, then, do DNA sequences from avian influenza become incorporated into human influenza?

Solution:
Pigs are infected by both avian and human influenza. The human and avian strains re-assort within pigs and the new viruses are then transmitted to humans.

APPLICATION QUESTIONS AND PROBLEMS

Section 7.2

*12.   John Smith is a pig farmer. For the past five years, Smith has been adding vitamins and low doses of antibiotics to his pig food; he says that these supplements enhance the growth of the pigs. Within the past year, however, several of his pigs died from infections of common bacteria, which failed to respond to large doses of antibiotics. Can you explain the increased rate of mortality due to infection in Smith's pigs? What advice might you offer Smith to prevent this problem in the future?

Solution:
By using low doses of antibiotics for five years, Farmer Smith selected for bacteria that are resistant to the antibiotics. The doses used killed sensitive bacteria but not moderately sensitive or slightly resistant bacteria. As time passed, only resistant bacteria remained in his pigs because any sensitive bacteria were eliminated by the low doses of antibiotics.

In the future, Farmer Smith can continue to use the vitamins, but should use the antibiotics only when a sick pig requires them. In this manner, he will not be selecting for antibiotic-resistant bacteria, and the chances of successful treatment of his sick pigs will be greater.

13.   In **Figure 7.8**, what do the red and blue parts of the DNA labeled by balloon 6 represent?

Solution:
The red part represents genes originally from the Y24 strain and the blue part represents genes that originated from the Y10 strain.

*14.   [Data Analysis Problem] Austin Taylor and Edward Adelberg isolated some new strains of Hfr cells that they then used to map several genes in *E. coli* by using interrupted conjugation (A. L. Taylor and E. A. Adelberg. 1960. *Genetics* 45:1233–1243). In one experiment, they mixed cells of Hfr strain AB-312, which were xyl^+ mtl^+ mal^+ met^+ and sensitive to phage T6, with F$^-$ strain AB-531, which was xyl^- mtl^- mal^- met^- and resistant to phage T6. The cells were allowed to undergo conjugation. At regular intervals, the researchers removed a sample of cells and interrupted conjugation by killing the Hfr cells with phage T6. The F$^-$ cells, which were resistant to phage T6, survived and were then tested for the presence of genes transferred from the Hfr strain. The results of this experiment are shown in the accompanying graph. On the basis of these data, give the order of the *xyl*, *mtl*, *mal*, and *met* genes on the bacterial chromosome and indicate the minimum distances between them.

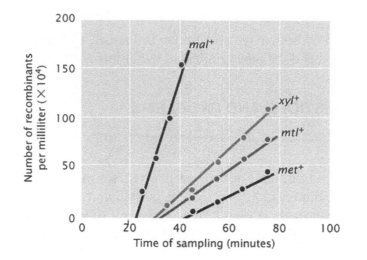

Solution:
The closer genes are to the F factor, the more quickly they will be transferred and more recombinants will be produced. The transfer process will occur in a linear fashion. By interrupting the mating process, the transfer will stop and the F⁻ strain will have received only genes carried on the piece of the Hfr strain's chromosome that entered the F⁻ cell prior to the disruption. From the graph, we can determine when the first recombinants for each marker were first identified and subsequently approximate the minutes that separate the different genetic markers.

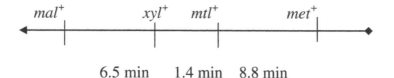

15. DNA from a strain of *Bacillus subtilis* with genotype $a^+ b^+ c^+ d^+ e^+$ is used to transform a strain with genotype $a^- b^- c^- d^- e^-$. Pairs of genes are checked for cotransformation and the following results are obtained:

Pair of genes	Cotransformation
a^+ and b^+	No
a^+ and c^+	No
a^+ and d^+	Yes
a^+ and e^+	Yes
b^+ and c^+	Yes
b^+ and d^+	No
b^+ and e^+	Yes
c^+ and d^+	No
c^+ and e^+	Yes
d^+ and e^+	No

On the basis of these results, what is the order of the genes on the bacterial chromosome?

Solution:
Only genes located near each other on the bacterial chromosome will be cotransformed together. However, by performing transformation experiments and screening for different pairs of cotransforming genes, a map of the gene order can be determined. Gene pairs that never result in cotransformation must be farther apart on the chromosome, whereas gene pairs that result in cotransformation are more closely linked. From the data, we see that gene a^+ cotransforms with both e^+ and d^+. However, genes d^+ and e^+ do not exhibit cotransformation, indicating that a^+ and e^+ are more closely linked than d^+ and e^+. Gene a^+ does not exhibit cotransformation with either gene b^+ or c^+, yet gene e^+ does. This indicates that gene e^+ is more closely linked to genes b^+ and c^+ than is gene a^+. The orientation of genes b^+ and c^+ relative to e^+ cannot be determined from the data provided.

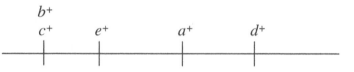

16. DNA from a bacterial strain that is his^+ leu^+ lac^+ is used to transform a strain that is his^- leu^- lac^-. The following percentages of cells were transformed:

Donor strain	Recipient strain	Genotype of transformed cells	Percentage of all cells
his^+ leu^+ lac^+	his^- leu^- lac^-	his^+ leu^+ lac^+	0.02
		his^+ leu^+ lac^-	0.00
		his^+ leu^- lac^+	2.00
		his^+ leu^- lac^-	4.00
		his^- leu^+ lac^+	0.10
		his^- leu^- lac^+	3.00
		his^- leu^+ lac^-	1.5

a. What conclusions can you draw about the order of these three genes on the chromosome?

Solution:
The percentages of cotransformation between his^+, leu^+, and lac^+ loci must be examined. Genes that cotransform more frequently will be closer together on the donor chromosome. Cotransformation between lac^+ and his^+ occurs in 2.02% of the transformed cells. By comparing this value with the cotransformation of lac^+ and leu^+ at 0.12% and the cotransformation of his^+ and leu^+ at 0.02%, we can see that lac^+ and his^+ cotransform more frequently. Therefore, lac^+ and his^+ must be more closely linked than the other gene pair combinations. Because lac^+ and leu^+ cotransform more frequently than his^+ and leu^+, leu^+ must be located closer to lac^+ than it is to his^+.

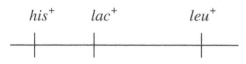

b. Which two genes are closest?

Solution:
From the cotransformation frequencies, we can predict that lac^+ and his^+ are the closest two genes.

*17. [Data Analysis Problem] Rollin Hotchkiss and Julius Marmur studied transformation in the bacterium *Streptococcus pneumoniae* (R. D. Hotchkiss and J. Marmur. 1954. *Proceedings of the National Academy of Sciences of the United States of America* 40:55– 60). They examined four mutations in this bacterium: penicillin resistance (P), streptomycin resistance (S), sulfanilamide resistance (F), and the ability to utilize mannitol (M). They extracted DNA from strains of bacteria with different combinations of different mutations and used this DNA to transform wild-type bacterial cells ($P^+ S^+ F^+ M^+$). The results from one of their transformation experiments are shown.

Donor DNA	Recipient DNA	Transformants	Percentage of all cells
$M S F$	$M^+ S^+ F^+$	$M^+ S F^+$	4.0
		$M^+ S^+ F$	4.0
		$M S^+ F^+$	2.6
		$M S F^+$	0.41
		$M^+ S F$	0.22
		$M S^+ F$	0.0058
		$M S F$	0.0071

a. Hotchkiss and Marmur noted that the percentage of cotransformation was higher than would be expected on a random basis. For example, the results show that the 2.6% of the cells were transformed into M and 4% were transformed into S. If the M and S traits were inherited independently, the expected probability of cotransformation of M and S ($M S$) would be $0.026 \times 0.04 = 0.001$, or 0.1%. However, they observed 0.41% $M S$ cotransformants, four times more than they expected. What accounts for the relatively high frequency of cotransformation of the traits they observed?

Solution:
It is likely that the M and S traits are linked or, in other words, they are located very close to each other on the *Streptococcus pneumoniae* chromosome. By being located close together on the chromosome, these markers are more likely to be cotransformed on a single fragment of DNA.

b. On the basis of the results, what conclusion can you make about the order of the M, S, and F genes on the bacterial chromosome?

Solution:
Because M and S cotransform quite frequently, they are likely close to each other on the chromosome as indicated above. S and F cotransform more frequently (0.22) than do M and F (0.0058). The transformation data suggests that S and F are located closer together than are M and F.

c. Why is the rate of cotransformation for all three genes ($M\,S\,F$) almost the same as the cotransformation of $M\,F$ alone?

Solution:
Genes M and F cotransform infrequently, which is likely due to the physical distance between them. Genes M and S are more closely linked on chromosome and the relative positions of M, S, and F make it likely that, if M and F are cotransformed on the same DNA molecule, then S will be cotransformed as well.

Section 7.3

*18. Two mutations that affect plaque morphology in phages (a^- and b^-) have been isolated. Phages carrying both mutations ($a^-\,b^-$) are mixed with wild-type phages ($a^+\,b^+$) and added to a culture of bacterial cells. Subsequent to infection and lysis, samples of the phage lysate are collected and cultured on bacterial cells. The following numbers of plaques are observed:

Plaque phenotype	Number
$a^+\,b^+$	2043
$a^+\,b^-$	320
$a^-\,b^+$	357
$a^+\,b^-$	2134

What is the frequency of recombination between the a and b genes?

Solution:
First, we must identify the progeny phage whose plaque phenotype is different from either of the infecting phage. The original infecting phages were wild-type ($a^+\,b^+$) and doubly mutant ($a\,b$). Any phages that give rise to the $a^+\,b^-$ plaque phenotype or the $a^-\,b^+$ plaque phenotype were produced by recombination between the two types of infecting phage particles.

Plaque phenotype	Number
$a^+\,b^+$	2043
$a^+\,b$	320 (recombinant)
$a^-\,b^+$	357 (recombinant)
$a^-\,b^-$	2134
Total plaques	4854

The frequency of recombination is calculated by dividing the total number of recombinant plaques by the total number of plaques (677/4854), which gives a frequency of 0.14, or 14%.

*19. [Data Analysis Problem] T. Miyake and M. Demerec examined proline-requiring mutations in the bacterium *Salmonella typhimurium* (T. Miyake and M. Demerec. 1960. *Genetics* 45:755–762). On the basis of complementation studies, they found four proline auxotrophs: *proA*, *proB*, *proC*, and *proD*. To determine whether *proA*, *proB*, *proC*, and *proD* loci were located close together on the bacterial chromosome, they conducted a transduction experiment. Bacterial strains that were *proC*⁺ and had mutations at *proA*, *proB*, or *proD* were used as donors. The donors were infected with bacteriophages, and progeny phages were allowed to infect recipient bacteria with genotype *proC⁻ proA⁺ proB⁺ proD⁺*. The recipient bacteria were then plated on a selective medium that allowed only *proC*⁺ bacteria to grow. After this, the *proC*⁺ transductants were plated on selective media to reveal their genotypes at the other three *pro* loci. The following results were obtained:

Donor genotype	Transductant genotype	Number
$proC^+ proA^- proB^+ proD^+$	$proC^+ proA^+ proB^+ proD^+$	2765
	$proC^+ proA^- proB^+ proD^+$	3
$proC^+ proA^+ proB^- proD^+$	$proC^+ proA^+ proB^+ proD^+$	1838
	$proC^+ proA^+ proB^- proD^+$	2
$proC^+ proA^+ proB^+ proD^-$	$proC^+ proA^+ proB^+ proD^+$	1166
	$proC^+ proA^+ proB^+ proD^-$	0

a. Why are there no *proC⁻* genotypes among the transductants?

Solution:
Transductants were initially screened for the presence of *proC*⁺. Thus, only *proC*⁺ transductants were identified.

b. Which genotypes represent single transductants and which represent cotransductants?

Solution:
The wild-type genotypes (*proC⁺ proA⁺ proB⁺ proD⁺*) represent single transductants of *proC*⁺. Both the *proC⁺ proA⁻ proB⁺ proD⁺* and *proC⁺ proA⁺ proB⁻ proD⁺* genotypes represent cotransductants of *proC⁺, proA⁻* and *proC⁺, proB⁻*.

c. Is there evidence that *proA*, *proB*, and *proD* are located close to *proC*? Explain your answer.

Solution:
From the data, it appears that both *proA* and *proB* are located close to *proC*. Both are capable of being cotransduced along with *proC*. The *proD* marker may be located at distance from *proC* so that it cannot contransduce with *proC*. However, the data is not conclusive.

*20. A geneticist isolates two mutations in a bacteriophage. One mutation causes clear plaques (*c*) and the other produces minute plaques (*m*). Previous mapping experiments have established that the genes responsible for these two mutations are 8 m.u. apart. The geneticist mixes phages with genotype $c^+ m^+$ and genotype $c^- m^-$ and uses the mixture to infect bacterial cells. She collects the progeny phages and cultures a sample of them on plated bacteria. A total of 1000 plaques are observed. What numbers of the different types of plaques ($c^+ m^+$, $c^- m^-$, $c^+ m^-$, $c^- m^+$) should she expect to see?

Solution:
We know that the two genes are 8 map units apart. These 8 map units correspond to a percent recombination between the two genes of 8%. When the geneticist mixes the two phages ($m^+ c^+ \times m^- c^-$), creating a double infection of the bacterial cell, she should expect the two types of recombinant plaque phenotypes, $m^+ c^-$ and $m^- c^+$, to comprise 8% of the progeny phage. The remaining 92% will be a combination of the wild-type phage and the doubly mutant phage.

Plaque phenotype	Expected number
$c^+ m^+$	460
$c^- m^-$	460
$c^+ m^-$	40 (recombinant)
$c^+ m^-$	40 (recombinant)
Total plaques	1000

*21. *E. coli* cells are simultaneously infected with two strains of phage λ. One strain has a mutant host range, is temperature sensitive, and produces clear plaques (genotype *h st c*); another strain carries the wild-type alleles (genotype $h^+ st^+ c^+$). Progeny phages are collected from the lysed cells and plated on bacteria. The numbers of different progeny phages are follows:

Progeny phage genotype			Number of plaques
h^+	c^+	st^+	321
h	c	st	338
h^+	c	st	26
h	c^+	st^+	30
h^+	c	st^+	106
h	c^+	st	110
h^+	c^+	st	5
h	c	st^+	6

a. Determine the order of the three genes on the phage chromosome.

Solution:
First, we need to identify the progeny phages that have genotypes similar to the parents and the progeny phages that have genotypes that differ from the parents. The parental genotypes are $h^+ c^+ st^+$ and *h c st*. Any genotype that differs from these two genotypes had to be generated by recombination. By comparing the genotype of the double-recombinant phage progeny with the nonrecombinants, we can predict the gene order.

Phage genotype	Number of progeny	Type
$h^+c^+ st^+$	321	Parental
$h\ c\ st$	338	Parental
$h^+\ c\ st$	26	Recombinant
$h\ c^+\ st^+$	30	Recombinant
$h^+\ c\ st^+$	106	Recombinant
$h\ c^+\ st$	110	Recombinant
$h^+\ c^+\ st$	5	Double-recombinant
$h\ c\ st^+$	6	Double-recombinant
Total	942	

$$h^+ \qquad\qquad st^+ \qquad\qquad c^+$$

b. Determine the map distances between the genes.

Solution:
The map distances can be calculated by determining the percent recombination between each gene pair. The double-recombinant progeny, $h^+ c^+$ and $h\ c$, appear to be parentals. However, this genotype was generated by a double-crossover event. To consider the double-crossover events, multiply the number of double-recombinant progeny by two.

$h^+ st^+$: $[(26 + 30 + 5 + 6)/942] \times 100\% = 7.1\%$, or 7.1 m.u.
$h^+ c^+$: $[(26 + 30 + 106 + 110 + 10 + 12)/942] \times 100\% = 31.2\%$, or 31.2 m.u.
$st^+ c^+$: $[(106 + 110 + 5 + 6)/942] \times 100\% = 24.1\%$ or 24.1 m.u.

$$h^+ \qquad\qquad st^+ \qquad\qquad c^+$$
$$\text{7.1 m.u.} \qquad \text{24.1 m.u.}$$

c. Determine the coefficient of coincidence and the interference (see pp. 136–137 in Chapter 5).

Solution:

$$\text{coefficient of coincidence} = \frac{\text{(observed number of double recombinants)}}{\text{(expected number of double recombinants)}}$$

coefficient of coincidence $= (6+5) / (0.071 \times 0.241 \times 942) = 0.68$

Interference $= 1 - \text{coefficient of coincidence} = 1 - 0.68 = 0.32$

22. A donor strain of bacteria with genes $a^+ b^+ c^+$ is infected with phages to map the donor chromosome with generalized transduction. The phage lysate from the bacterial cells is collected and used to infect a second strain of bacteria that are $a^- b^- c^-$. Bacteria with the a^+ gene are selected, and the percentage of cells with cotransduced b^+ and c^+ genes are recorded.

Donor	Recipient	Selected gene	Cells with cotransduced gene (%)
$a^+\ b^+\ c^+$	$a^-\ b^-\ c^-$	a^+	$25\ b^+$
		a^+	$3\ c^+$

Is the b or c gene closer to a? Explain your reasoning.

Solution:
The gene b^+ cotransduces more frequently with a^+, the selective marker, than does c^+. Because genes that are closer together on the donor bacterial chromosome cotransduce more frequently, we can see that b^+ is closer to a^+.

23. For the H1N1 influenza virus shown at the bottom of **Figure 7.29**, viruses from which organism contributed the most RNA to the virus?

Solution:
Swine

CHALLENGE QUESTION

Section 7.2

24. A group of genetics students mixes two auxotrophic strains of bacteria: one is $leu^+\ trp^+$ $his^-\ met^-$ and the other is $leu^-\ trp^-\ his^+\ met^+$. After mixing the two strains, they plate the bacteria on minimal medium and observe a few prototrophic colonies ($leu^+\ trp^+\ his^+\ met^+$). They assume that some gene transfer has occurred between the two strains. How can they determine whether the transfer of genes is due to conjugation, transduction, or transformation?

Solution:
Conjugation requires the direct contact of the donor bacterial strain and the recipient If the transfer does not occur when the bacteria are kept physically separate, then conjugation is not the likely pathway. Another test would be to conduct interrupted mating experiments (assuming that one of the bacterial strains is an Hfr strain) to see if the transfer of the different markers is time dependent, which is also indicative of conjugation.

If the transfer occurs by transformation, then extraction of DNA from either strain and exposure of the other strain to the extracted DNA should result in the transfer of the DNA molecules. By selecting one of the mutations as a selective marker and measuring cotransformation frequencies between the selective marker and the other genes individually, the frequency of the transfer will hint toward the mechanism of transfer.

Finally, if the transfer is by transduction, then, by exposing the one cell type to extracted DNA from the other cell type, transfer of the genes would not be expected. Potential cotransduction frequencies could be measured similarly to the cotransformation frequency. Also, the presence of plaques might be evident.

Chapter Eight: DNA: The Chemical Nature of the Gene

COMPREHENSION QUESTIONS

Section 8.1

1.   What four general characteristics must the genetic material possess?

Solution:
(1) The genetic material must contain complex information.
(2) The genetic material must replicate or be replicated faithfully.
(3) The genetic material must have the capacity to vary or mutate to generate diversity.
(4) The genetic material must encode the phenotype or have the ability to code for traits.

Section 8.2

2.   What is transformation? How did Avery and his colleagues demonstrate that the transforming principle is DNA?

Solution:
Transformation occurs when a transforming material (or DNA) genetically alters the bacterium that absorbs the transforming material. Avery and his colleagues demonstrated that DNA is the transforming material by using enzymes that destroyed the different classes of biological molecules. Enzymes that destroyed proteins or nucleic acids had no effect on the activity of the transforming material. However, enzymes that destroyed DNA eliminated the biological activity of the transforming material. Avery and his colleagues were also able to isolate the transforming material and demonstrate that it had chemical properties similar to DNA.

3.   How did Hershey and Chase show that DNA is passed to new phages in phage reproduction?

Solution:
Hershey and Chase used ^{32}P-labeled phages to infect bacteria. The progeny phage released from these bacteria emitted radioactivity from ^{32}P. The presence of the ^{32}P in the progeny phage indicated that the infecting phages had passed DNA on to the progeny phage.

Section 8.3

4.   Draw and identify the three parts of a DNA nucleotide.

Solution:
The three parts of a DNA nucleotide are phosphate, deoxyribose sugar, and a nitrogenous base.

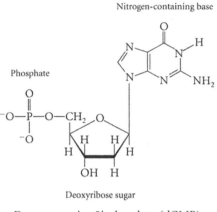

Deoxyguanosine 5'-phosphate (dGMP)

5.   How does an RNA nucleotide differ from a DNA nucleotide?

Solution:
DNA nucleotides, or deoxyribonucleotides, have a deoxyribose sugar that lacks an oxygen molecule at the 2′ carbon of the sugar molecule. Ribonucleotides, or RNA nucleotides, have a ribose sugar with an oxygen linked to the 2′ carbon of the sugar molecule. Ribonucleotides may contain the nitrogenous base uracil, but not thymine. DNA nucleotides contain thymine, but not uracil.

6.   Draw a short segment of a single DNA polynucleotide strand, including at least three nucleotides. Indicate the polarity of the strand by labeling the 5′ end and the 3′ end.

Solution:

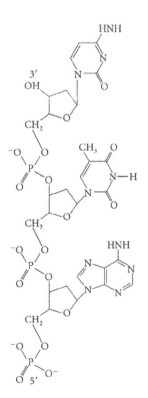

7.   What are some of the important genetic implications of the DNA structure?

Solution:
Referring back to Question 1, the structure of DNA gives insight into the three fundamental genetic processes. The Watson and Crick model suggests that the genetic information or instructions are encoded in the nucleotide sequences. The complementary polynucleotide strands indicate how faithful replication of the genetic material is possible. Finally, the arrangement of the nucleotides is such that they specify the primary structure or amino acid sequence of protein molecules.

8.   What are the three major pathways of information flow within the cell?

Solution:
Replication, transcription, and translation are the components of the central dogma of molecular biology.

Section 8.4

9.   How does supercoiling arise? What is the difference between positive and negative supercoiling?

Solution:
Supercoiling arises from overwinding (**positive** supercoiling) or underwinding (**negative** supercoiling) of the DNA double helix; from a lack of free ends, as in circular DNA molecules; when the ends of the DNA molecule are bound to proteins that prevent them from rotating about each other.

10.   What functions does supercoiling serve for the cell?

Solution:
Supercoiling compacts the DNA. Negative supercoiling helps to unwind the DNA duplex for replication and transcription.

11.   What are some differences between euchromatin and heterochromatin?

Solution:
Euchromatin undergoes regular cycles of condensation during mitosis and decondensation during interphase, whereas heterochromatin remains highly condensed throughout the cell cycle, except transiently during replication. Nearly all transcription takes place in euchromatic regions, with little or no transcription of heterochromatin.

12.   Describe the composition and structure of the nucleosome.

Solution:
The nucleosome core particle contains two molecules each of histones H2A, H2B, H3, and H4, which form a protein core with 145–147 bp of DNA wound around the core. A chromatosome contains the nucleosome core and a molecule of histone H1.

13.   Describe in steps how the double helix of DNA, which is 2-nm wide, gives rise to a chromosome that is 700-nm wide.

Solution:
DNA is first packaged into nucleosomes; the nucleosomes are packed to form a 30-nm fiber. The 30-nm fiber forms a series of loops that pack to form a 250-nm fiber, which in turn coils to form a 700-nm chromatid.

14.   What are epigenetic changes?

Solution:
Epigenetic changes are changes in gene expression that are passed on to cells or future generations, but do not involve alternation of the nucleotide sequence. Epigenetic changes are brought about by altering DNA structure, such as methylation of the DNA, or altering chromatin structure.

Section 8.5

15.   Describe the function and molecular structure of a telomere.

Solution:
Telomeres are the stable ends of the linear chromosomes in eukaryotes. They prevent degradation by exonucleases and prevent joining of the ends. Telomeres also enable the replication of the ends of the chromosome. Telomeric DNA sequences consist of repeats of a simple sequence, usually in the form of $5'–C_n(A \text{ or } T)_m$.

16.   Describe the different classes of DNA sequence variation that exist in eukaryotes.

Solution:
Unique-sequence DNA, present in only one or a few copies per haploid genome, represents most of the protein coding sequences, plus a great deal of sequences with unknown function.

Moderately repetitive sequences, a few hundred to a few thousand base pairs long, are present in up to several thousand copies per haploid genome. Some moderately repetitive DNA consists of functional genes that code for rRNAs and tRNAs, but most is made up of transposable elements and remnants of transposable elements.

Highly repetitive DNA, or satellite DNA, consists of clusters of tandem repeats of short (often less than 10 bp) sequences present in hundreds of thousands to millions of copies per haploid genome.

APPLICATION QUESTIONS AND PROBLEMS

Introduction

17. The introduction to this chapter, which describes the sequencing of 4000-year-old DNA, emphasizes DNA's extreme stability. What aspects of DNA's structure contribute to the stability of the molecule? Why is RNA less stable than DNA?

Solution:
Several aspects contribute to the stability of the DNA molecule. The relatively strong phosphodiester linkages connect the nucleotides of a given strand of DNA. The helical nature of the double-stranded DNA molecule results in the negatively charged phosphates of each strand being arranged to the outside and away from each other. The complementary nature of the nitrogenous bases of the nucleotides helps hold the two strands of polynucleotides together. The stacking interactions of the bases, which allow for any base to follow another in a given strand, also play a major role in holding the two strands together. Finally, the ability of DNA to have local variations in secondary structure contributes to its stability.

RNA nucleotides or ribonucleotides contain an extra oxygen at the 2′ carbon of the ribose sugar. This extra oxygen at each nucleotide makes RNA a less stable molecule.

18. Match the scientists with the discoveries listed.
    a. Kossel
    b. Watson and Crick
    c. Levene
    d. Miescher
    e. Hershey and Chase
    f. Avery, MacLeod, and McCarty
    g. Griffith
    h. Franklin and Wilkins
    i. Chargaff

Solution:
    __h__ Took X-ray diffraction pictures used in determining the structure of DNA.
    __a__ Determined that DNA contains nitrogenous bases.
    __e__ Identified DNA as the genetic material in bacteriophage.
    __i__ Discovered regularity in the ratios of different bases in DNA.
    __f__ Determined that DNA is responsible for transformation in bacteria.
    __b__ Worked out the helical structure of DNA by building models.
    __c__ Discovered that DNA consists of repeating nucleotides.

    d   Determined that DNA is acidic and high in phosphorus.

    g   Demonstrated that heat-killed material from bacteria could genetically transform live bacteria.

Section 8.2

*19. A student mixes some heat-killed-type IIS *Streptococcus pneumoniae* bacteria with live type IIR bacteria and injects the mixture into a mouse. The mouse develops pneumonia and dies. The student recovers some type IIS bacteria from the dead mouse. If this is the only experiment conducted by the student, has the student demonstrated that transformation has taken place? What other explanations might explain the presence of the type IIS bacteria in the dead mouse?

Solution:
No, the student has not demonstrated that transformation has taken place. A single mutation could convert the IIR strain into the virulent IIS strain. Thus, the student cannot determine whether the conversion from IIR to IIS is due to transformation or due to a mutation.

Also, the student has not demonstrated that the heat was sufficient to kill all the IIS bacteria. A second useful control experiment would have been to inject the heat-killed IIS into mice and see if any of the IIS bacteria survived the heat treatment.

20. Predict what would have happened if Griffith had mixed some heat-killed type IIIS bacteria and some heat-killed type IIR bacteria and injected this mixture into a mouse. Would the mouse have contracted pneumonia and died? Explain why or why not.

Solution:
No, the mouse would not have contracted pneumonia and died. Although the mouse would have received IIIS DNA, which codes for virulent *Streptococcus pneumoniae*, there are no live bacteria for this DNA to transform. Live bacteria are required for pneumonia to develop.

Section 8.3

*21. DNA molecules of different sizes are often separated using a technique called electrophoresis (see Chapter 14). With this technique, DNA molecules are placed in a gel, an electrical current is applied to the gel, and the DNA molecules migrate toward the positive (+) pole of the current. What aspect of its structure causes a DNA molecule to migrate toward the positive pole?

Solution:
The phosphate backbone of DNA molecules typically carries a negative charge, thus making the DNA molecules attractive to the positive pole of the current.

*22. [Data Analysis Problem] Erwin Chargaff collected data on the proportions of nucleotide bases from the DNA of a variety of different organisms and tissues (E. Chargaff, in *The Nucleic Acids: Chemistry and Biology*, vol. 1, E. Chargaff and J. N. Davidson, Eds. New York: Academic Press, 1955). Data from the DNA of several organisms analyzed by Chargaff are shown here.

Organism and tissue	A	G	C	T
		Percent		
Sheep thymus	29.3	21.4	21.0	28.3
Pig liver	29.4	20.5	20.5	29.7
Human thymus	30.9	19.9	19.8	29.4
Rat bone marrow	28.6	21.4	20.4	28.4
Hen erythrocytes	28.8	20.5	21.5	29.2
Yeast	31.7	18.3	17.4	32.6
E. coli	26.0	24.9	25.2	23.9
Human sperm	30.9	19.1	18.4	31.6
Salmon sperm	29.7	20.8	20.4	29.1
Herring sperm	27.8	22.1	20.7	27.5

a. For each organism, compute the ratio of $(A + G)/(T + C)$ and the ratio of $(A + T)/(C + G)$.

Solution:

Organism	$(A + G)/(T + C)$	$(A + T)/(C + G)$
Sheep thymus	1.03	1.36
Pig liver	0.99	1.44
Human thymus	1.03	1.52
Rat bone marrow	1.02	1.36
Hen erythrocytes	0.97	1.38
Yeast	1.00	1.80
E. coli	1.04	1.00
Human sperm	1.00	1.67
Salmon sperm	1.02	1.43
Herring sperm	1.04	1.29

b. Are these ratios constant or do they vary among the organisms? Explain why.

Solution:
The $(A + G)/(T + C)$ ratios of ~1.0 is constant for these organisms. Each of them has a double-stranded genome. The percentage of purines should be equal to the percentage of pyrimidines for double-stranded DNA, which means that $(A + G) = (T + C)$. The $(A + T)/(C + G)$ ratios are not constant. The number of A + T base pairs and C + G base pairs are unique to each organism and can vary among the different organisms.

c. Is the (A + G)/(T + C) ratio different for the sperm samples? Would you expect it to be? Why or why not?

Solution:
The (A+G)/(T+C) ratio is about the same, as should be expected. As in part **b**, the percentage of purines should equal the percentage of pyrimidines.

23. [Data Analysis Problem] Boris Magasanik collected data on the amounts of the bases of RNA isolated from a number of sources (shown here), expressed relative to a value of 10 for adenine (B. Magasanik, in *The Nucleic Acids: Chemistry and Biology*, vol. 1, E. Chargaff and J. N. Davidson, Eds. New York: Academic Press, 1955).

Organism and tissue	A	G	C	U
		Percent		
Rat liver nuclei	10	14.8	14.3	12.9
Rabbit liver nuclei	10	13.6	13.1	14.0
Cat brain	10	14.7	12.0	9.5
Carp muscle	10	21.0	19.0	11.0
Yeast	10	12.0	8.0	9.8

a. For each organism, compute the ratio of (A + G)/(U + C).

Solution:

Organism and tissue	(A + G)/(U + C)
Rat liver nuclei	0.91
Rabbit liver nuclei	0.87
Cat brain	1.15
Carp muscle	1.03
Yeast	1.24

b. How do these ratios compare with the (A + G)/(T + C) ratio found in DNA (see Problem 22)? Explain.

Solution
The ratios are not as similar to each other or as close to the value of 1.0 as found for the (A + G)/(T + C) ratio in DNA. Many RNA molecules are single-stranded and do not have large regions of complementary sequences as we would expect to find in DNA.

24. Which of the following relations or ratios would be true for a double-stranded DNA molecule?

Solution:
A double-stranded DNA molecule will contain equal percentages of A and T nucleotides and equal percentages of G and C nucleotides. The combined percentage of A and T bases added to the combined percentage of the G and C bases should equal 100.

a. $A + T = G + C$ No
b. $A + G = T + C$ Yes
c. $A + C = G + T$ Yes
d. $\dfrac{A+T}{C+G} = 1.0$ No
e. $\dfrac{A+G}{C+T} = 1.0$ Yes
f. $\dfrac{A}{C} = \dfrac{G}{T}$ No
g. $\dfrac{A}{G} = \dfrac{T}{C}$ Yes
h. $\dfrac{A}{T} = \dfrac{G}{C}$ Yes

*25. If a double-stranded DNA molecule is 15% thymine, what are the percentages of all the other bases?

Solution:
The percentage of thymine (15%) should be approximately equal to the percentage of adenine (15%). The remaining percentage of DNA bases will consist of cytosine and guanine bases (100% – 15% – 15% = 70%); these should be in equal amounts (70%/2 = 35%). Therefore, the percentages of each of the other bases if the thymine content is 15% are adenine = 15%; guanine = 35%; and cytosine = 35%.

26. [Data Analysis Problem] Heinz Shuster collected the following data on the base composition of the ribgrass virus (H. Shuster, in *The Nucleic Acids: Chemistry and Biology*, vol. 3, E. Chargaff and J. N. Davidson, Eds. New York: Academic Press, 1955). On the basis of this information, is the hereditary information of the ribgrass virus RNA or DNA? Is it likely to be single stranded or double stranded?

	A	G	C	T	U
			Percent		
Ribgrass virus	29.3	25.8	18.0	0.0	27.0

Solution:
Most likely, the ribgrass viral genome is a single-stranded RNA. The presence of uracil indicates that the viral genome is RNA. For the molecule to be double-stranded RNA, we would predict equal percentages of adenine and uracil bases and equal percentages of guanine and cytosine bases. Neither the percentages of adenine and uracil bases nor the percentages of guanine and cytosine bases are equal, indicating that the viral genome is likely single stranded.

*27. For entertainment on a Friday night, a genetics professor proposed that his children diagram a polynucleotide strand of DNA. Having learned about DNA in preschool, his five-year-old daughter was able to draw a polynucleotide strand, but she made a few mistakes. Sarah's diagram contains at least 10 mistakes.

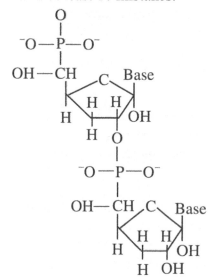

a. Make a list of all the mistakes in the structure of this DNA polynucleotide strand.

Solution:
(1) Neither 5′-carbon atom of the two sugars is directly linked to phosphorus.
(2) Neither 5′-carbon atom of the two sugars has an OH group attached.
(3) Neither sugar molecule has oxygen in its ring structure between the 1′- and 4′-carbon atoms.
(4) In both sugars, the 2′-carbon atom has an –OH group attached, which the 2′-carbon atom of deoxyribonucleotides does not have.
(5) A hydrogen atom is attached at the 3′ position in both sugars, rather than an –OH group.
(6) An –OH group is attached at the 1′-carbon of both sugars rather than a hydrogen atom.

b. Draw the correct structure for the polynucleotide strand.

Solution:

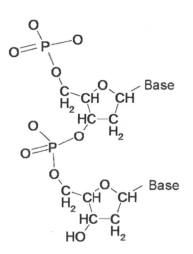

28.   One nucleotide strand of DNA molecule has the base sequence illustrated below.

5′—ATTGCTACGG—3′

Give the base sequence and label the 5′ and 3′ ends of the complementary DNA nucleotide strand.

Solution:
Answer: 3′—TAACGATGCC—5′

*29.   Chapter 1 considered the theory of the inheritance of acquired characteristics and noted that this theory is no longer accepted. Is the central dogma consistent with the theory of the inheritance of acquired characteristics? Why or why not?

Solution:
No. The flow of information predicted by the central dogma is from DNA to RNA to protein. An exception to the central dogma is reverse transcription, whereby RNA encodes DNA. However, biologists currently do not know of a process in which the flow of information is from proteins to DNA, which is required by the theory of the inheritance of acquired characteristics.

30.   Which of the processes of information transfer illustrated in **Figure 8.14** are required for the T2 phage reproduction illustrated in **Figure 8.4**?

Solution:
DNA replication, transcription, and translation.

Section 8.5

*31.   Compare and contrast prokaryotic and eukaryotic chromosomes. How are they alike and how do they differ?

Solution:
Prokaryotic chromosomes are usually circular, whereas eukaryotic chromosomes are linear. Prokaryotic chromosomes generally contain the entire genome, whereas each eukaryotic chromosome has only a portion of the genome. The eukaryotic genome is divided into multiple chromosomes. Prokaryotic chromosomes are generally much smaller than eukaryotic chromosomes and have only a single origin of DNA replication, whereas eukaryotic chromosomes contain multiple origins of DNA replication. Prokaryotic chromosomes are typically condensed into nucleoids, which have loops of DNA compacted into a dense body. Eukaryotic chromosomes contain DNA packaged into nucleosomes, which are further coiled and packaged into structures of successively higher order. The condensation state of eukaryotic chromosomes varies with the cell cycle.

*32. A diploid human cell contains approximately 6.4 billion base pairs of DNA.
   a. How many nucleosomes are present in such a cell? (Assume that the linker DNA encompasses 40 bp.)

   Solution:
   Given that each nucleosome contains about 140 bp of DNA tightly associated with the core histone octamer, another 20 bp associated with histone H1, and 40 bp in the linker region, then one nucleosome occurs for every 200 bp of DNA.

   6.4×10^9 bp divided by 2×10^2 bp/nucleosome = 3.2×10^7 nucleosomes (32 million).

   b. How many histone proteins are complexed with this DNA?

   Solution:
   Each nucleosome contains two of each of the following histones: H2A, H2B, H3, and H4. A nucleosome plus one molecule of histone H1 constitute the chromatosome. Therefore, nine histone protein molecules occur for every nucleosome.

   3.2×10^7 nucleosomes $\times$ 9 histones = 2.9×10^8 molecules of histones are complexed to 6.4 billion bp of DNA.

CHALLENGE QUESTIONS

Section 8.1

*33. Suppose that an automated, unmanned probe is sent into deep space to search for extraterrestrial life. After wandering for many light-years among the far reaches of the universe, this probe arrives on a distant planet and detects life. The chemical composition of life on this planet is completely different from that of life on Earth, and its genetic material is not composed of nucleic acids. What predictions can you make about the chemical properties of the genetic material on this planet?

   Solution:
   Although the chemical composition of the genetic material may be different from that of DNA, it more than likely will have similar properties to those of DNA. As stated in this chapter, the genetic material must contain complex information, replicate or be replicated faithfully, and encode the phenotype. Even if the material on the planet is not DNA, it must meet these criteria. Additionally, the genetic material must be stable.

Section 8.2

34. How might ^{32}P and ^{35}S be used to demonstrate that the transforming principle is DNA? Briefly outline an experiment that would show that DNA, rather than protein, is the transforming principle.

Solution:
The first step would be to label the DNA and proteins of the donor bacteria cells with ^{32}P and ^{35}S. The DNA could be labeled by growing a culture of bacteria in the presence of ^{32}P. The cells as they replicate ultimately will incorporate radioactive phosphorus into their DNA. A second culture of bacteria should be grown in the presence of ^{35}S, which ultimately will be incorporated into proteins.

Material from each culture should be used to transform bacteria cells that previously had not been exposed to the radioactive isotopes. Transformed cells (or colonies) that would be identified by the acquisition of a new phenotype should contain low levels of the radioactive material due to the uptake of the labeled molecules. If the transforming material were protein, then cells transformed by the material from the ^{35}S exposed bacterial cultures would also contain ^{35}S. If the transforming material were DNA, then the cells transformed by the material from the ^{32}P exposed bacterial cultures would also contain ^{32}P.

Section 8.3

35.   Researchers have proposed that early life on Earth used RNA as its source of genetic information and that DNA eventually replaced RNA as the source of genetic information. What aspects of DNA structure might make it better suited than RNA to be the genetic material?

Solution:
The carrier of genetic information must be stable so that the genetic information is faithfully transmitted from one generation to the next. The 3′ hydroxyl group on the ribose sugar of RNA makes RNA more reactive than DNA, which lacks a free 3′ hydroxyl group on its deoxyribose sugar. Because it is single stranded, RNA can assume a number of secondary structures and a number of different functions, decreasing its stability. The double-stranded nature of DNA stabilizes DNA and makes it more suitable as the carrier of genetic information. If mistakes occur in one strand, the complementary strand can serve as a template for corrections.

*36.   Imagine that you are a student in Alfred Hershey and Martha Chase's laboratory in the late 1940s. You are given five test tubes containing *E. coli* bacteria infected with T2 bacteriophages that have been labeled with either ^{32}P or ^{35}S. Unfortunately, you forgot to indicate which tubes are labeled with ^{32}P and which with ^{35}S. You place the contents of each tube in a blender and turn it on for a few seconds to shear off the protein coats. You then centrifuge the contents to separate the protein coats and the cells, check for the presence of radioactivity, and obtain the results shown here. Which tubes contained *E. coli* infected with ^{32}P-labeled phage? Explain your answer.

Tube number	Radioactivity present in
1	Cells
2	Protein coats
3	Protein coats
4	Cells
5	Cells

Solution:

Tubes 1, 4, and 5. The DNA of the bacteriophage contains phosphorus and the protein contains sulfur. When the bacteriophages infect the cell, they inject their DNA into the cell, but the phage protein coats stay on the surface of the cell. The protein coats are sheared off in the blender, while the cells with the DNA pellet at the bottom of the tube. Thus, cells infected with ^{35}S-labeled bacteriophage will have radioactivity associated with the protein coats, whereas those cells infected with ^{32}P-labeled bacteriophage will have radioactivity associated with the cells.

Chapter Nine: DNA Replication and Recombination

COMPREHENSION QUESTIONS

Section 9.2

1.   What is semiconservative replication?

Solution:
In semiconservative replication, the original two strands of the double helix serve as templates for new strands of DNA. When replication is complete, two double-stranded DNA molecules will be present. Each will consist of one original template strand and one newly synthesized strand that is complementary to the template.

2.   How did Meselson and Stahl demonstrate that replication in *E. coli* takes place in a semiconservative manner?

Solution:
Meselson and Stahl grew *E. coli* cells in a medium containing the heavy isotope of nitrogen (^{15}N) for several generations. The ^{15}N was incorporated in the DNA of the *E. coli* cells. The *E. coli* cells were then switched to a medium containing the common form of nitrogen (^{14}N) and allowed to proceed through a few cycles of cellular generations. Samples of the bacteria were removed at each cellular generation.

Using equilibrium density gradient centrifugation, Meselson and Stahl were able to distinguish DNAs that contained only ^{15}N from DNAs that contained only ^{14}N or a mixture of ^{15}N and ^{14}N because DNAs containing the ^{15}N isotope are "heavier." The more ^{15}N in a DNA fragment, the lower the fragment will sink. DNA from cells grown in the ^{15}N medium produced only a single band at the expected position during centrifugation. After one round of replication in the ^{14}N medium, one band was present following centrifugation, but the band was located at an intermediate position between that of a DNA band containing only ^{15}N and that of a DNA band containing only ^{14}N. After two rounds of replication, two bands of DNA were present. One band was located at an intermediate position between that of a DNA band containing only ^{15}N and that of a DNA band containing only ^{14}N, whereas the other band was at a position expected for DNA containing only ^{14}N.

These results were consistent with the predictions of semiconservative replication and incompatible with the predictions of conservative and dispersive replication.

3.   Draw a molecule of DNA undergoing replication. On your drawing, identify (1) origin, (2) polarity (5′ and 3′ ends) of all template strands and newly synthesized strands, (3) leading and lagging strands, (4) Okazaki fragments, and (5) location of primers.

Solution:

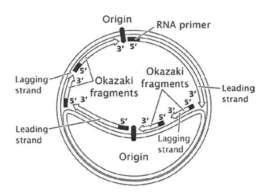

4. Draw a molecule of DNA undergoing eukaryotic linear replication. On your drawing, identify (1) origin, (2) polarity (5′ and 3′ ends) of all template and newly synthesized strands, (3) leading and lagging strands, (4) Okazaki fragments, and (5) location of primers.

    Solution:

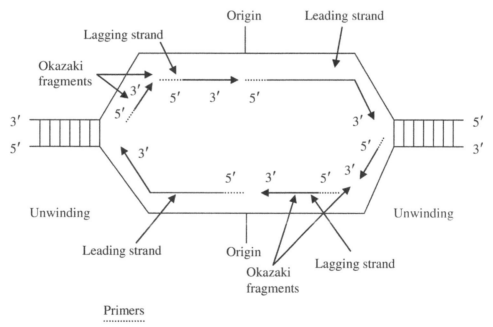

5. What are three major requirements of replication?

    Solution:
    (1) A single-stranded DNA template
    (2) Nucleotide substrates for synthesis of the new polynucleotide strand
    (3) Enzymes and other proteins associated with replication to assemble the nucleotide substrates into a new DNA molecule

6.   What substrates are used in the DNA synthesis reaction?

Solution:
The substrates for DNA synthesis are the four types of deoxyribonucleoside triphosphates: deoxyadenosine triphosphate, deoxyguanosine triphosphate, deoxycytosine triphosphate, and deoxythymidine triphosphate.

Section 9.3

7.   List the different proteins and enzymes taking part in bacterial replication. Give the function of each in the replication process.

Solution:
DNA polymerase III is the primary replication polymerase. It elongates a new nucleotide strand from the 3′–OH of the primer.

DNA polymerase I removes the RNA nucleotides of the primers and replaces them with DNA nucleotides.

DNA ligase connects Okazaki fragments by sealing nicks in the sugar phosphate backbone.

DNA primase synthesizes the RNA primers that provide the 3′–OH group needed for DNA polymerase III to initiate DNA synthesis.

DNA helicase unwinds the double helix by breaking the hydrogen bonding between the two strands at the replication fork.

DNA gyrase reduces DNA supercoiling and torsional strain that is created ahead of the replication fork by making double-stranded breaks in the DNA and passing another segment of the helix through the break before resealing it. Gyrase is also called topoisomerase II.

Initiator proteins bind to the replication origin and unwind short regions of DNA.

Single-stranded binding protein (SSB protein) stabilizes single-stranded DNA prior to replication by binding to it, thus preventing the DNA from pairing with complementary sequences.

8.   What similarities and differences exist in the enzymatic activities of DNA polymerases I and III? What is the function of each type of DNA polymerase in bacterial cells?

Solution:
Each of the DNA polymerases has a 5′ to 3′ polymerase activity. They differ in their exonuclease activities. DNA polymerase I has a 3′ to 5′ as well as a 5′ to 3′ exonuclease activity. DNA polymerase III has only a 3′ to 5′ exonuclease activity.
(1) DNA polymerase I carries out proofreading. It also removes and replaces the RNA primers used to initiate DNA synthesis.
(2) DNA polymerase III is the primary replication enzyme and also has a proofreading function in replication.

9.   Why is primase required for replication?

Solution:
Primase, an RNA polymerase, synthesizes the short RNA molecules, or primers, that provide a 3′–OH to which DNA polymerase can attach deoxyribonucleotides in replication initiation.

10.   Why is DNA gyrase necessary for replication?

Solution:
DNA synthesis relies on a single-stranded template; thus, double-stranded DNA molecules must be unwound prior to replication. During DNA unwinding by DNA helicase, tension builds up ahead of the separation (supercoiling). DNA gyrase (also referred to as topoisomerase) reduces supercoiling (relaxes tension) which builds up during DNA unwinding, preventing DNA breakage.

11.   What three mechanisms ensure the accuracy of replication in bacteria?

Solution:
(1) Highly accurate nucleotide selection by the DNA polymerases when pairing bases.
(2) The proofreading function of DNA polymerase, which removes incorrectly inserted bases.
(3) A mismatch repair apparatus that repairs mistakes after replication is complete.

12.   How does replication licensing ensure that DNA is replicated only once at each origin per eukaryotic cell cycle?

Solution:
Only replication origins to which replication licensing factor (RPF) has bound can undergo initiation. Shortly after the completion of mitosis, RPF binds the origin during G_1 and is removed by the replication machinery during S phase.

13.   In what ways is eukaryotic replication similar to bacterial replication, and in what ways is it different?

Solution:
Eukaryotic and bacterial replication of DNA share some basic principles:
(1) Replication is semiconservative.
(2) Replication origins serve as starting points for replication.
(3) A short segment of RNA called a primer provides a 3′-OH for DNA polymerases to begin the synthesis of the new strands.
(4) Synthesis is in the 5′ to 3′ direction.
(5) The template strand is read in the 3′-to-5′ direction.
(6) Deoxyribonucleoside triphosphates are the substrates.

(7) Replication is continuous on the leading strand and discontinuous on the lagging strand.

Eukaryotic DNA replication differs from bacterial replication in the following ways:
(1) There are multiple origins of replications per chromosome.
(2) Several different DNA polymerases have different functions.
(3) Immediately after DNA replication, nucleosomes are assembled.

14.   What is the end-replication problem? Why, in the absence of telomerase, do the ends of linear chromosomes get progressively shorter each time the DNA is replicated?

Solution:
For DNA polymerases to work, they need the presence of a 3'-OH group to which to add a nucleotide. At the ends of the chromosomes when the RNA primer is removed, there is no adjacent 3'-OH group to which to add a nucleotide, thus no nucleotides are added, which leaves a gap at the end of the chromosome. Telomerase can extend the single-stranded protruding end by pairing with the overhanging 3' end of the DNA and adding a repeated sequence of nucleotides. In the absence of telomerase, DNA polymerase will be unable to add nucleotides to the end of the strand. After multiple rounds of replication without a functional telomerase, the chromosome will become progressively shorter.

15.   Outline in words and pictures how telomeres at the ends of eukaryotic chromosomes are replicated.

Solution:
Telomeres are replicated by the enzyme telomerase. Telomerase, a ribonucleoprotein, consists of protein and an RNA molecule that is complementary to the 3' end of the DNA of a eukaryotic chromosome. The RNA molecule also serves as a template for the addition of nucleotides to the 3' end. After the 3' end has been extended, the 5' end of the DNA can be extended as well, possibly by lagging strand synthesis of a DNA polymerase using the extended 3' end as a template.

DNA replication of the linear eukaryotic chromosomes generates a 3' overhang. Part of the RNA sequence within telomerase is complementary to the overhang.

```
        Telomerase

        5'–CAACCCCAA–
```

```
    5'–CCCCAA ——————————————
  3'–GGGGTTGGGGTT——————————————
                Toward centromere  ⟶
```

Telomerase RNA sequence pairs with the 3′ overhang and serves as a template for the addition of DNA nucleotides to the 3′ end of the DNA molecule, which serves to extend the 3′ end of the chromosome.

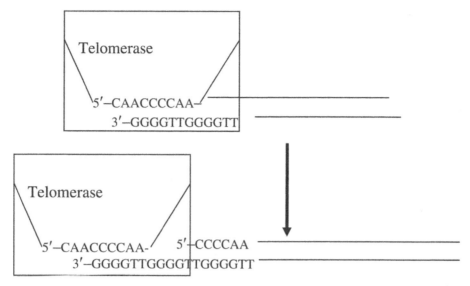

Additional nucleotides to the 5′ end are added by DNA synthesis using a DNA polymerase with priming by primase.

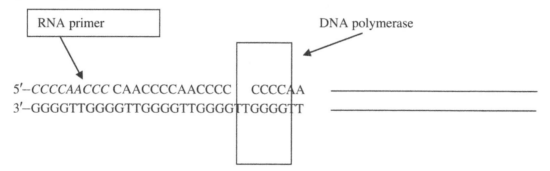

APPLICATION QUESTIONS AND PROBLEMS

Section 9.2

16.   Suppose that a future scientist explores a distant planet and discovers a novel form of double-stranded nucleic acid. When this nucleic acid is exposed to DNA polymerases from *E. coli*, replication takes place continuously on both strands. What conclusion can you make about the structure of this novel nucleic acid?

Solution:
Each strand of the novel double-stranded nucleic acid must be oriented parallel to the other, as opposed to the antiparallel nature of earthly double-stranded DNA. Replication by *E. coli* DNA polymerases can proceed continuously only in a 5′ to 3′ direction, which requires the template to be read in a 3′ to 5′ direction. If replication is continuous on both strands, the two strands must have the same direction and be parallel.

17.   Phosphorus is required to synthesize the deoxyribonucleoside triphosphates used in DNA replication. A geneticist grows some *E. coli* in a medium containing nonradioactive phosphorus for many generations. A sample of the bacteria is then transferred to a medium that contains a radioactive isotope of phosphorus (^{32}P). Samples of the bacteria are removed immediately after the transfer and after one and two rounds of replication. Assume that newly synthesized DNA contains ^{32}P and that the original DNA contains nonradioactive phosphorus. What will be the distribution of radioactivity in the DNA of the bacteria in each sample? Will radioactivity be detected in neither strand, one strand, or both strands of the DNA?

Solution:
In the initial sample removed immediately after transfer, no ^{32}P should be incorporated into the DNA because replication in the medium containing ^{32}P has not yet occurred. After one round of replication in the ^{32}P containing medium, one strand of each newly synthesized DNA molecule will contain ^{32}P, whereas the other strand will contain only nonradioactive phosphorus. After two rounds of replication in the ^{32}P containing medium, 50% of the DNA molecules will have ^{32}P in both strands, whereas the remaining 50% will contain ^{32}P in one strand and nonradioactive phosphorus in the other strand.

*18. A line of mouse cells is grown for many generations in a medium with ^{15}N. Cells in G_1 are then switched to a new medium that contains ^{14}N. Draw a pair of homologous chromosomes from these cells at the following stages, showing the two strands of DNA molecules found in the chromosomes. Use different colors to represent strands with ^{14}N and ^{15}N. (See Chapter 2 for review of the stages of the cell cycle and meiosis.)

   a.   Cells in G_1, before switching to medium with ^{14}N

        Solution:

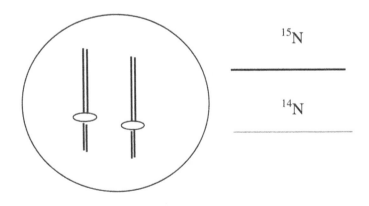

b. Cells in G₂, after switching to medium with ^{14}N

   Solution:

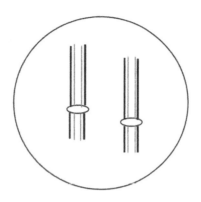

c. Cells in anaphase of mitosis, after switching to medium with ^{14}N

   Solution:

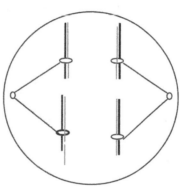

d. Cells in metaphase I of meiosis, after switching to medium with ^{14}N

   Solution:

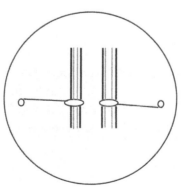

e. Cells in anaphase II of meiosis, after switching to medium with ^{14}N

Solution:

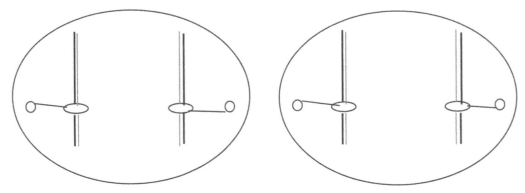

*19. A circular molecule of DNA contains 1 million base pairs. If the rate of DNA synthesis at a replication fork is 100,000 nucleotides per minute, how much time will theta replication require to completely replicate the molecule, assuming that theta replication is bidirectional?

Solution:
In bidirectional replication there are two replication forks, each proceeding at a rate of 100,000 nucleotides per minute. Therefore, it would require 5 minutes for the circular DNA molecule to be replicated by bidirectional replication because each fork could synthesize 500,000 nucleotides (5 minutes × 100,000 nucleotides per minute) within the time period. Because rolling-circle replication is unidirectional and thus has only one replication fork, 10 minutes will be required to replicate the entire circular molecule.

20. A bacterium synthesizes DNA at each replication fork at a rate of 1000 nucleotides per second. If this bacterium completely replicates its circular chromosome by theta replication in 30 minutes, how many base pairs of DNA will its chromosome contain?

Solution:
Each replication complex is synthesizing DNA at each fork at a rate of 1000 nucleotides per second. So, for each second, 2000 nucleotides are being synthesized by both forks (1000 nucleotides/second × 2 forks = 2000 nucleotides/second), or 120,000 nucleotides per minute. If the bacterium requires 30 minutes to replicate its chromosome, then the size of the chromosome is 3,600,000 nucleotide base pairs (120,000 nucleotides/minute × 30 minutes = 3,600,000).

Section 9.3

21. In **Figure 9.7**, which is the leading strand and which is the lagging strand?

Solution:
The top red strand is lagging and the bottom red strand is leading.

*22. The following diagram represents a DNA molecule that is undergoing replication. Draw in the strands of newly synthesized DNA and identify (a) through (d):
   a. Polarity of newly synthesized strands
   b. Leading and lagging strands
   c. Okazaki fragments
   d. RNA primers

Solution:

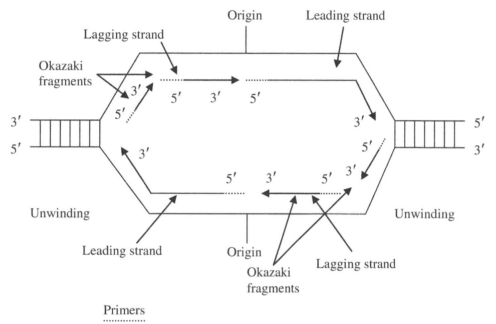

*23. What would be the effect on DNA replication of mutations that destroyed each of the following activities in DNA polymerase I?
   a. $3' \rightarrow 5'$ exonuclease activity

   Solution:
   The $3' \rightarrow 5'$ exonuclease activity is important for proofreading newly synthesized DNA. If the activity is nonfunctional, then the fidelity of replication by DNA polymerase I will decrease, resulting in more misincorporated bases in the DNA.

   b. $5' \rightarrow 3'$ exonuclease activity

   Solution:
   Loss of the $5' \rightarrow 3'$ exonuclease activity would result in the RNA primers used to initiate replication not being removed by DNA polymerase I.

c. $5' \rightarrow 3'$ polymerase activity

Solution:
DNA polymerase I would be unable to synthesize new DNA strands if the $5' \rightarrow 3'$ polymerase activity was destroyed. RNA primers could be removed by DNA polymerase I using the $5' \rightarrow 3'$ exonuclease activity, but new DNA sequences could not be added in their place by DNA polymerase I.

24.   How would DNA replication be affected in a bacterial cell that is lacking DNA gyrase?

Solution:
DNA gyrase or topoisomerase II reduces the positive supercoiling or torsional strain that develops ahead of the replication fork due to the unwinding of the double helix. If the topoisomerase activity was lacking, then the torsional strain would continue to increase, making it more difficult to unwind the double helix. Ultimately, the increasing strain would lead to an inhibition of the replication fork movement.

25.   Arrange the following components of replication in the order in which they first act in the replication process: ligase, DNA polymerase I, helicase, gyrase, primase, single-strand-binding protein, initiator proteins.

Solution:
Initiator protein, helicase, gyrase, single-strand binding proteins, primase, DNA polymerase I, ligase.

Section 9.4

26.   A number of scientists who study cancer treatment have become interested in telomerase. Why? How might cancer-drug therapies that target telomerase work?

Solution:
Telomerase is an enzyme that functions in cells that undergo continuous cell division and may play a role in the lack of cellular aging in these cells. Cells that lack telomerase exhibit progressive shortening of the chromosomal ends or telomeres. This shortening leads to unstable chromosomes and ultimately to programmed cell death. Many tumor cells also express telomerase, which may assist these cells in becoming immortal. If the telomerase activity in these cancer cells could be inhibited, then cell division might be halted in the cancer cells thus controlling cancer cell growth. Chemically modified antisense RNAs or DNA oligonucleotides complementary to the telomerase RNA sequence might block the telomerase activity by base pairing with the telomerase RNA, making it unavailable as a template. A second strategy would be to target the DNA synthesis activity of the telomerase protein preventing the telomere DNA from being synthesized.

*27.   The enzyme telomerase is part protein and part RNA. What would be the most likely effect of deleting the gene that encodes the RNA part of telomerase? How would the function of telomerase be affected?

Solution:
The RNA portion of telomerase is needed to provide the template for synthesizing complementary DNA telomere sequences at the ends of the chromosomes. A large deletion would affect how the telomeres are synthesized at the ends of the chromosomes by telomerase and could potentially eliminate telomere synthesis.

28.   [Data Analysis Problem] Dyskeratosis congenita (DKC) is a rare genetic disorder characterized by abnormal fingernails and skin pigmentation, the formation of white patches on the tongue and cheek, and progressive failure of the bone marrow. An autosomal dominant form of DKC results from mutations in the gene that encodes the RNA component of telomerase. Tom Vulliamy and his colleagues examined a series of families with autosomal dominant DKC (T. Vulliamy et al. 2004. *Nature Genetics* 36:447–449). They observed that the median age of onset of DKC in parents was 37 years, whereas the median age of onset in the children of affected parents was 14.5 years. Thus, DKC in these families arose at progressively younger ages in successive generations, a phenomenon known as anticipation. The researchers measured the telomere length of members of these families; the measurements are given in the accompanying table. Telomeres normally shorten with age, so telomere length was adjusted for age. Note that the age-adjusted telomere length of all members of these families is negative, indicating that their telomeres are shorter than normal. For age-adjusted telomere length, the more negative the number, the shorter the telomere.

Parent telomere length	Child telomere length
−4.7	−6.1
	−6.6
	−6.0
−3.9	−0.6
−1.4	−2.2
−5.2	−5.4
−2.2	−3.6
−4.4	−2.0
−4.3	−6.8
−5.0	−3.8
−5.3	−6.4
−0.6	−2.5
−1.3	−5.1
	−3.9
−4.2	−5.9

a. How does the telomere length of the parents compare with the telomere length of the children? (Hint: Calculate the average telomere length of all parents and the average telomere length of all children.)

Solution:
In general, the telomere lengths of the parents are longer than the telomere lengths of the children (with few exceptions).

b. Explain why the telomeres of people with DKC are shorter than normal.

Solution:
Persons with DKC possess an allele that codes for a mutant form of the RNA component of telomerase. The RNA component serves as the template for telomere synthesis and is necessary for the synthesis of telomeres of appropriate lengths.

c. Explain why DKC arises at an earlier age in subsequent generations.

Solution:
The defective RNA component results in telomeres that are shortened. For persons in subsequent generations who inherit the DKC, it is most likely that these person also inherited shorter telomeres from the parent possessing the DKC allele as evidenced from the above data.

Once telomeres become too short, cells normally undergo cell death (apoptosis), which likely leads to the symptoms of DKC. In subsequent generations, telomeres may reach that critical length earlier, leading to an earlier onset of DKC.

Section 9.5

29. An person is heterozygous at two loci (*Ee Ff*) and the two genes are in repulsion (see p. 124 in Chapter 5). Assume that single-strand breaks and branch migration occur at the positions shown below. Using different colors to represent the two homologous chromosomes, draw the noncrossover recombinant and crossover recombinant DNA molecules that will result from homologous recombination. (Hint: See **Figure 9.16**.)

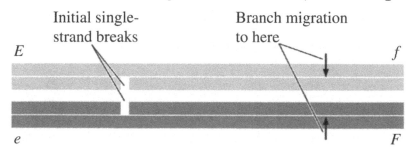

Solution:
Noncrossover recombinants

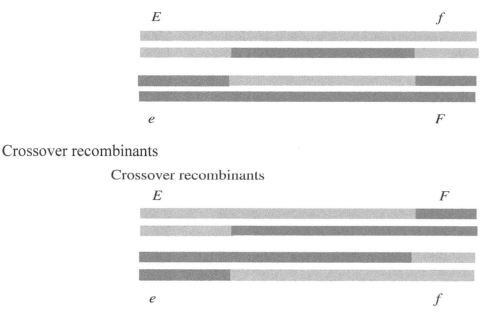

Crossover recombinants

CHALLENGE QUESTIONS

Section 9.3

30.   A conditional mutation is one that expresses its mutant phenotype only under certain conditions (the restrictive conditions) and expresses the normal phenotype under other conditions (the permissive conditions). One type of conditional mutation is a temperature-sensitive mutation, which expresses the mutant phenotype only at certain temperatures.

Strains of *E. coli* have been isolated that contain temperature-sensitive mutations in the genes encoding different components of the replication machinery. In each of these strains, the protein produced by the mutated gene is nonfunctional under the restrictive conditions. These strains are grown under permissive conditions and then abruptly switched to the restrictive condition. After one round of replication under the restrictive condition, the DNA from each strain is isolated and analyzed. What characteristics would you expect to see in the DNA isolated from each strain with a temperature-sensitive mutation in its gene that encodes in the following proteins?

a. DNA ligase

Solution:
DNA ligase is required to seal the nicks left after DNA polymerase I removes the RNA primers used to begin DNA synthesis. If DNA ligase is not functioning, multiple nicks in the lagging strand will be expected and the Okazaki fragments will not be joined. However, DNA replication will take place.

b. DNA polymerase I

Solution:
RNA primers are removed by DNA polymerase I. After one round of replication, the DNA molecule would still contain the RNA nucleotide primers since DNA polymerase I is not functioning. However, replication of the DNA will take place.

c. DNA polymerase III

Solution:
No replication would be expected. DNA polymerase III is the primary replication enzyme. If it is not functioning, then neither lagging- nor leading-strand DNA synthesis will take place.

d. Primase

Solution:
No replication would be expected. Primase synthesizes the short RNA molecules that act as primers for DNA synthesis. If the RNA primers are not synthesized, then no free 3′–OH will be available for DNA polymerase III to attach DNA nucleotides. Thus, DNA synthesis will not take place.

e. Initiator proteins

Solution:
Initiator proteins bind to the oriC and unwind the DNA, allowing for the binding of DNA helicase and single-stranded binding proteins to the DNA. If these proteins are unable to bind, then DNA replication initiation will not occur, so DNA synthesis will not take place.

31.   [Data Analysis Problem] DNA topoisomerases play important roles in DNA replication and supercoiling (see Chapter 8). These enzymes are also the targets for certain anticancer drugs. Eric Nelson and his colleagues studied m-AMSA, an anticancer compound that acts on topisomerases. They found that m-AMSA stabilizes an intermediate produced in the course of the topoisomerase's action. The intermediate consisted of the topoisomerase bound to the broken ends of the DNA (E. M. Nelson, K. M. Tewey, and L. F. Liu. 1984. *Proceedings of the National Academy of Sciences of the United States of America* 81:1361–1365). Breaks in DNA that are produced by anticancer compounds such as m-AMSA inhibit the replication of the cellular DNA and thus stop cancer cells from proliferating. Explain how m-AMSA and other anticancer agents that target topoisomerase enzymes taking part in replication might lead to DNA breaks and chromosome rearrangements.

Solution:
Compounds such as m-AMSA prevent topoisomerases from completing their functions. Type II topoisomerases (such as gyrase) function to reduce torsional strain generated ahead of the replication fork. The enzymes reduce the strain by generating a double-stranded break in a segment of the DNA and passing another segment of the DNA through the break, which is then followed by the resealing of the broken ends. The anticancer agent m-AMSA and related substances inhibit the function of topoisomerases by stabilizing the topoisomerase on the 5' ends of the double-stranded break. This prevents the other segment from passing through and prevents resealing of the break. Essentially, the compounds generate double-stranded breaks in the DNA due to the action of topoisomerase. Breaks in the DNA molecule may stimulate recombination and repair enzymes leading to chromosomal rearrangements.

Chapter Ten: From DNA to Proteins: Transcription and RNA Processing

COMPREHENSION QUESTIONS

Section 10.1

1. Draw an RNA nucleotide and a DNA nucleotide, highlighting the differences. How is the structure of RNA similar to that of DNA? How is it different?

    Solution:
    RNA and DNA are polymers of nucleotides that are held together by phosphodiester bonds. An RNA nucleotide contains ribose, whereas a DNA nucleotide contains deoxyribose. RNA contains uracil but not thymine. DNA contains thymine but not uracil. RNA is typically single stranded, even though RNA molecules can pair with other complementary sequences. DNA molecules are almost always double stranded.

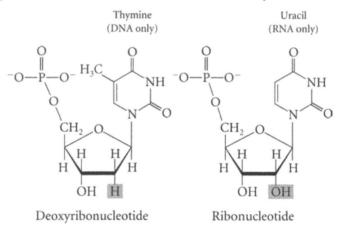

2. What are the major classes of cellular RNA?

    Solution:
    Cellular RNA molecules are made up of six classes:
    (1) Ribosomal RNA, or rRNA, is found in the cytoplasm.
    (2) Transfer RNA, or tRNA, is found in the cytoplasm.
    (3) Messenger RNA, or mRNA, is found in the cytoplasm (however, pre-mRNA is found only in the nucleus).
    (4) Small nuclear RNA, or snRNA, is found in the nucleus as part of riboproteins called snrps.
    (5) Small nucleolar RNA, snoRNA, is found in the nucleus.
    (6) Small cytoplasmic RNA, or scRNA, is found in the cytoplasm.

Section 10.2

3.   What parts of DNA make up a transcription unit? Draw a typical bacterial transcription unit and identify its parts.

Solution:

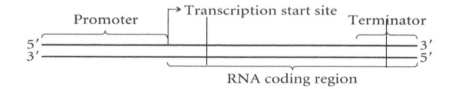

4.   What is the substrate for RNA synthesis? How is this substrate modified and joined together to produce an RNA molecule?

Solution:
Four ribonucleoside triphosphates serve as the substrate for RNA synthesis: adenosine triphosphate, guanosine triphosphate, cytosine triphosphate, and uridine monophosphate. The enzyme RNA polymerase uses a DNA polynucleotide strand as a template to synthesize a complementary RNA polynucleotide strand. The nucleotides are added to the RNA polynucleotide strand, one at time, at the 3′–OH of the RNA molecule. As each nucleoside triphosphate is added to the growing polynucleotide chain, two phosphates are removed from the 5′ end of the nucleotide. The remaining phosphate is linked to the 3′–OH of the RNA molecule to form the phosphodiester bond.

Section 10.3

5.   What are the three basic stages of transcription? Describe what takes place at each stage.

Solution:
   (1) Initiation: Transcription proteins assemble at the promoter to form the basal transcription apparatus and begin synthesis of RNA.
   (2) Elongation: RNA polymerase moves along the DNA template in a 3′ to 5′ direction unwinding the DNA and synthesizing RNA in a 5′ to 3′ direction.
   (3) Termination: Synthesis of RNA is terminated, and the RNA molecule separates from the DNA template.

6.   How are transcription and replication similar and how are they different?

Solution:
Common characteristics of transcription and replication:
(1) Utilize a DNA template
(2) Synthesize molecules in a 5′ to 3′ direction
(3) Synthesize molecules that are antiparallel and complementary to the template
(4) Use nucleotide triphosphates as substrates
(5) Involve complexes of proteins and enzymes necessary for catalysis

Unique characteristics of transcription:
(1) Unidirectional synthesis of only a single strand of nucleic acid
(2) Initiation does not require a primer
(3) Subject to numerous regulatory mechanisms
(4) Each gene is transcribed separately
Unique characteristics of replication:
(1) Bidirectional synthesis of two strands of nucleic acid
(2) Initiates from replication origins

Section 10.5

7.   What are the three principal elements in mRNA sequences in bacterial cells?

Solution:
(1) The 5′ untranslated region, which contains the Shine-Dalgarno sequence
(2) The protein-encoding region
(3) The 3′ untranslated region

8.   What is the function of the Shine-Dalgarno consensus sequence?

Solution:
The Shine-Dalgarno consensus sequence functions as the ribosome-binding site on the mRNA molecule.

9.   What is the 5′ cap?

Solution:
The 5′ end of eukaryotic mRNA is modified by the addition of the 5′ cap. The cap consists of an extra guanine nucleotide (linked 5′ to 5′ to the mRNA molecule) that is methylated at position 7 of the base and at adjacent bases whose sugars are methylated at the 2′–OH.

10.   What is the function of the spliceosome?

Solution:
The spliceosome consists of five snRNPs. The splicing of pre-mRNA nuclear introns takes place within the spliceosome.

11.   What is alternative splicing? How does it lead to the production of multiple proteins from a single gene?

Solution:
Most genes in eukaryotes include one or more introns that must be removed from the initial product of transcription, which is called a pre-mRNA molecule. As the introns are removed, the exons (which carry the information needed to encode the polypeptide product) are spliced together. Alternative splicing allows for pre-mRNA molecules from a single gene to be spliced

in multiple ways so that a particular stretch of nucleotides could be treated as an intron or as part of an exon, depending on the cell type or stage of development. Therefore, different mRNA molecules carrying different coding regions can be produced from a single gene, allowing for a single gene to encode more than one polypeptide.

12.   Summarize the different types of processing that can take place in pre-mRNA.

Solution:
Several modifications to pre-mRNA take place to produce mature mRNA.
(1) Addition of the 5′ cap to the 5′ end of the pre-mRNA
(2) Cleavage of the 3′ end of a site downstream of the AAUAAA consensus sequence of the last exon
(3) Addition of the poly(A) tail to the 3′ end of the mRNA immediately following cleavage
(4) Removal of the introns (splicing)

13.   Briefly describe the structure of tRNAs.

Solution:
tRNA molecules are single-stranded polynucleotides that include A, G, C, and U nucleotides, as well as dozens of additional modified bases. Because some of the stretches of nucleotides within the tRNA are complementary to each other, the single strand folds back on itself to form a complex, cloverleaf structure stabilized by hydrogen bonds. This structure folds further to form an L-shaped tertiary structure. The 3′ end of a tRNA molecule includes the sequence 5′-CCA-3′, where the amino acid attaches to the tRNA. At the other end of the molecule is a group of three adjacent nucleotides that constitute the anticodon, which allows the tRNA to match up with a codon on the mRNA molecule being translated.

14.   What is the origin of small interfering RNAs and microRNAs? What do these RNA molecules do in the cell?

Solution:
The siRNAs originate from the cleavage of mRNAs, RNA transposons, and RNA viruses by the enzyme Dicer. Dicer may produce multiple siRNAs from a single double-stranded RNA molecule. The double-stranded RNA molecule may occur due to the formation of hairpins or by duplexes between different RNA molecules. The miRNAs arise from the cleavage of individual RNA molecules that are distinct from other genes. The enzyme Dicer cleaves these RNA molecules that have formed small hairpins. A single miRNA is produced from a single RNA molecule.

Both siRNAs and microRNAs silence gene expression through a process called RNA interference. Both function by shutting off gene expression of a cell's own genes or by shutting off expression of genes from the invading foreign genes of viruses or tranposons. The microRNAs typically silence genes that are different from those from which the microRNAs are transcribed. However, the siRNAs usually silence genes from which they are transcribed.

APPLICATION QUESTIONS AND PROBLEMS

Section 10.1

*15. An RNA molecule has the following percentages of bases: A = 23%, U = 42%, C = 21%, and G = 14%.

    a. Is this RNA single-stranded or double-stranded? How can you tell?

    Solution:
    The RNA molecule is likely to be single-stranded. If the molecule were double-stranded, we would expect nearly equal percentages of adenine and uracil and equal percentages of guanine and cytosine.

    b. What would be the percentages of bases in the template strand of the DNA that contains the gene for this RNA?

    Solution:
    Because the DNA template strand is complementary to the RNA molecule, we would expect equal percentages of bases in the DNA complementary to the RNA bases. Therefore, in the DNA we would expect A = 42%, T = 23%, C = 14%, and G = 21%.

Section 10.2

*16. The following diagram represents DNA that is part of the RNA-coding sequence of a transcription unit. The bottom strand is the template strand. Give the sequence found on the RNA molecule transcribed from this DNA and identify the 5′ and 3′ ends of the RNA.
        5′–ATAGGCGATGCCA–3′
        3′–TATCCGCTACGGT–5′ ← Template strand

    Solution:
    The RNA molecule would be complementary to the template strand, contain uracil, and be synthesized in an antiparallel fashion. The sequence would be:
        5′–AUAGGCGAUGCCA–3′
    The RNA strand contains the same sequence as the nontemplate DNA strand, except that the RNA strand contains uracil in place of thymine.

17. For the RNA molecule shown in **Figure 10.1a**, write out the sequence of bases on the template and nontemplate strands of DNA from which this RNA is transcribed. Label the 5′ and 3′ ends of each strand.

    Solution:
    Template strand: 5′—GTCA—3′; Nontemplate strand: 5′—TGAC—3′

18. The following sequence of nucleotides is found in a single-stranded DNA template:
ATTGCCAGATCATCCCAATAGAT
Assume that RNA polymerase proceeds along this template from left to right.

   a. Which end of the DNA template is 5′ and which end is 3′?

   Solution:
   RNA is synthesized in a 5′ to 3′ direction by RNA polymerase, which reads the DNA template in a 3′ to 5′ direction. So, if the polymerase is moving from left to right on the template then the 3′ end must be on the left and the 5′ end on the right.
   3′–A T T G C C A G A T C A T C C C A A T A G A T–5′

   b. Give the sequence and identify the 5′ and 3′ ends of the RNA copied from this template.

   Solution:
   5′–U A A C G G U C U A G U A G G G U U A U C U A–3′

19. List at least five properties that DNA polymerases and RNA polymerases have in common. List at least three differences.

   Solution:
   Similarities: (1) Both use DNA templates, (2) DNA templates are read in the 3′ to 5′ direction, (3) the complementary strand is synthesized in a 5′ to 3′ direction, which is antiparallel to the template, (4) both use triphosphates as substrates, and (5) their actions are enhanced by accessory proteins.

   Differences: (1) RNA polymerases use ribonucleoside triphosphates as substrates, whereas DNA polymerases use deoxyribonucleoside triphophates; (2) DNA polymerases require a primer that provides an available 3′–OH group where synthesis begins, whereas RNA polymerases do not require primers to begin synthesis; and (3) RNA polymerases synthesize a copy off only one of the DNA strands, whereas DNA polymerases can synthesize copies off both strands.

20. Most RNA molecules have three phosphate groups at the 5′ end, but DNA molecules never do. Explain this difference.

   Solution:
   During initiation of DNA replication, DNA nucleotide triphosphates must be attached to a 3′–OH of a RNA molecule by DNA polymerase. This process removes the terminal two phosphates of the nucleotides. If the RNA molecule is subsequently removed, then a single phosphate would remain at the 5′ end of the DNA molecule. RNA polymerase does not require the 3′–OH to initiate synthesis of RNA molecules. Therefore, the 5′ end of a RNA molecule will retain all three of the phosphates from the original nucleotide triphosphate substrate.

21.   A strain of bacteria possesses a temperature-sensitive mutation in the gene that encodes the sigma factor. The mutant bacteria produce a sigma factor that is unable to bind to RNA polymerase at elevated temperatures. What effect will this mutation have on the process of transcription when the bacteria are raised at elevated temperatures?

Solution:
The binding of the sigma factor to the RNA polymerase core enzyme forms the RNA polymerase holoenzyme. Only the holoenzyme binds to the promoter.
Without the sigma factor, RNA polymerase will be unable to bind the promoter and transcription initiation will not take place. Any RNA polymerase that has completed transcription initiation and has begun elongation will complete transcription because the sigma factor is not needed for elongation. However, no further initiation will be possible at the elevated temperature.

22.   On **Figure 10.5**, indicate the locations of the promoters and terminators for genes *a, b,* and *c.*

Solution:

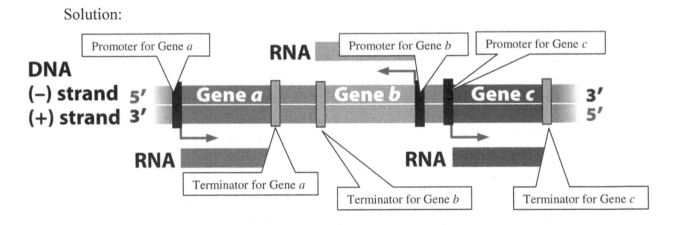

23.   The following diagram represents a transcription unit on a DNA molecule. Assume that this DNA molecule is from a bacterial cell. Label the approximate locations of the promoter and terminator for this transcription unit.

Solution:

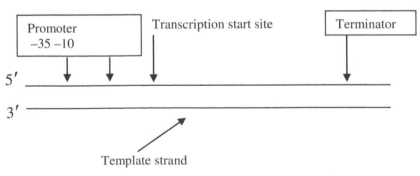

Section 10.3

24.   Provide the consensus sequence for the *first three* actual sequences shown in **Figure 10.9**.

Solution:
5′—TAYARNNG—3′

*25.   Write the consensus sequence for the following set of nucleotide sequences.

AGGAGTT
AGCTATT
TGCAATA
ACGAAAA
TCCTAAT
TGCAATT

Solution:
The consensus sequence is identified by determining which nucleotide is used most frequently at each position. For the two nucleotides that occur at an equal frequency at the first position, both are listed at that position in the sequence and identified by a slash mark:

T/AGAATT

*26.   What would be the most likely effect of a mutation at the following locations in an *E. coli* gene?

a. −8

Solution:
A mutation at the −8 position would probably affect the −10 consensus sequence (TATAAT), which is centered on position −10. This consensus sequence is necessary for binding of RNA polymerase. A mutation there would most likely decrease the rate of transcription.

b. −35

Solution:
A mutation in the −35 region could affect the binding of the sigma factor to the promoter. Mutations in the consensus sequences of the promoter region typically reduce or inhibit the rate of transcription.

c. −20

Solution:
The −20 region is located between the consensus sequences of an *E. coli* promoter. Although the holoenzyme may cover the site, it is unlikely that a mutation will have any effect on transcription.

d. Start site of transcription

Solution:
A mutation in the start site would have little effect on the rate of transcription. However, the RNA sequence may be altered which may ultimately disrupt the functionality of the transcript or the amino acid sequence of the polypeptide chain.

Section 10.4

*27. Duchenne muscular dystrophy is caused by a mutation in a gene that encompasses more than 2 million nucleotides and specifies a protein called dystrophin. However, less than 1% of the gene actually encodes the amino acids in the dystrophin protein. On the basis of what you now know about gene structure and RNA processing in eukaryotic cells, provide a possible explanation for the large size of the dystrophin gene.

Solution:
The large size of the gene that encodes dystrophin is likely due to the presence of many intervening sequences, or introns, within the coding region of the gene. Excision of the introns through RNA splicing yields the mature mRNA that encodes the dystrophin protein.

28.   What would be the most likely effect of moving the AAUAAA consensus sequence shown in **Figure 10.19** ten nucleotides upstream?

Solution:
The cleavage site would also be moved ten nucleotides upstream, likely resulting in a shorter mRNA.

29.   Suppose that a mutation occurs in the middle of a large intron of a gene encoding a protein. What will the most likely effect of the mutation be on the amino acid sequence of that protein? Explain your answer.

Solution:
Because introns are removed prior to translation, an intron mutation would have little effect on a protein's amino acid sequence unless the mutation occurred within the 5′ splice site, the 3′ splice site, or the branch point. If mutations within these sequences altered splicing such that a new exon is recognized by the spliceosome, then the mature mRNA would be altered, thus altering the amino acid sequence of the protein. The result could be a protein with additional amino acid sequence. Or, possibly, the inclusion of the new exon that was previously intron could introduce a stop codon that stops translation prematurely. If a mutation in the new exon or inclusion of the new exon induced a frameshift, the reading frame and the amino acid sequence would be altered from that point onward in the protein.

30. A geneticist isolates a gene that contains eight exons. He then isolates the mature mRNA produced by this gene. After making the DNA single stranded, he mixes the single-stranded DNA and RNA. Some of the single-stranded DNA hybridizes (pairs) with the complementary mRNA. Draw a picture of what the DNA–RNA hybrids will look like under the electron microscope.

Solution:

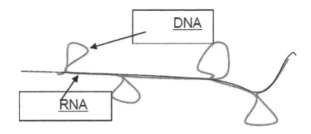

Section 10.5

*31. Draw a typical eukaryotic gene and the pre-mRNA and mRNA derived from it. Assume that the gene contains three exons. Identify the following items and, for each item, give a brief description of its function.

Solution:

a. 5′ untranslated region
The 5′ untranslated region lies upstream of the translation start site. In bacteria, the ribosome binding site or Shine-Dalgarno sequence is found within the 5′ untranslated region. However, eukaryotic mRNA does not have the equivalent sequence, and a eukaryotic ribosome binds at the 5′ cap of the mRNA molecule.

b. Promoter
The promoter is the DNA sequence that the transcription apparatus recognizes and binds to initiate transcription.

c. AAUAAA consensus sequence
The AAUAAA consensus sequence lies downstream of the coding region of the gene. It determines the location of the 3′ cleavage site in the pre-mRNA molecule.

d. Transcription start site
The transcription start site begins the coding region of the gene and is located 25 to 30 nucleotides downstream of the TATA box.

e. 3′ untranslated region
The 3′ untranslated region is a sequence of nucleotides at the 3′ end of the mRNA that is not translated into proteins. However, it does affect the translation of the mRNA molecule as well as the stability of the mRNA.

f. Introns

Introns are noncoding sequences of DNA that intervene within coding regions of a gene.

g. Exons

Exons are coding regions of a gene.

h. Poly(A) tail

A poly(A) tail is added to the 3′ end of the pre-mRNA; it affects mRNA stability and the binding of the ribosome to the mRNA.

i. 5′ cap

The 5′ cap functions in the initiation of translation and mRNA stability.

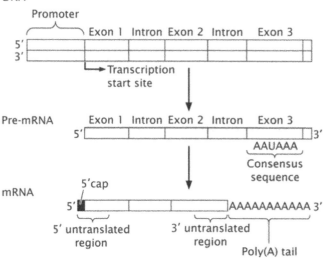

32.   How would the deletion of the Shine-Dalgarno sequence affect a bacterial mRNA?

Solution:
In bacteria, the small ribosomal subunit binds to the Shine-Dalgarno sequence to initiate translation. If the Shine-Dalgarno sequence is deleted, then translation initiation cannot take place, preventing protein synthesis.

33.   A geneticist discovers that two different proteins are encoded by the same gene. One protein has 56 amino acids, and the other has 82 amino acids. Provide a possible explanation for how the same gene can encode both of these proteins.

Solution:
The pre-mRNA molecules transcribed from the gene are likely processed by alternative processing pathways. Two possible mechanisms that could have produced the two different proteins from the same pre-mRNA are alternative splicing or multiple 3′ cleavage sites in the pre-mRNA. The cleavage of the pre-mRNA molecule at different 3′ cleavage sites would produce alternatively processed mRNA molecules that differ in size.

Translation from each of the alternative mRNAs would produce proteins containing different numbers of amino acids.

Alternative splicing of the pre-mRNA could produce different mature mRNAs, each containing a different number of exons. Again, translation from each alternatively spliced mRNA would generate proteins that differ in the number of amino acids contained.

34. [Data Analysis Problem] In the early 1990s, Carolyn Napoli and her colleagues were attempting to genetically engineer a variety of petunias with dark purple petals by introducing numerous copies of a gene that encodes purple petals (C. Napoli, C. Lemieux, and R. Jorgensen. 1990. *Plant Cell* 2:279–289). Their thinking was that extra copies of the gene would cause more purple pigment to be produced and would result in a petunia with an even darker hue of purple. However, much to their surprise, many of the plants carrying extra copies of the purple gene were completely white or had only patches of color. Molecular analysis revealed that the level of the mRNA produced by the purple gene was reduced 50-fold in the engineered plants compared with levels of mRNA in wild-type plants. Somehow, the introduction of extra copies of the purple gene silenced both the introduced copies and the plant's own purple genes. Provide a possible explanation for how the introduction of numerous copies of the purple gene silenced all copies of the purple gene.

Solution:
The overexpression of the purple gene mRNA led potentially to the formation of double-stranded regions by these RNA molecules because of areas of homology within the mRNAs being produced. These double-stranded molecules stimulated RNA silencing mechanisms or the RNA-Induced-Silencing Complex (RISC) leading to rapid degradation of the mRNA molecules. The result would be a reduction in translation of the protein needed for the production of purple petals and the phenotypic loss of pigmentation.

CHALLENGE QUESTIONS

Section 10.3

35. Many genes in both bacteria and eukaryotes contain numerous sequences that potentially cause pauses in or premature terminations of transcription. Nevertheless, the transcription of these genes within a cell normally produces multiple RNA molecules thousands of nucleotides long without pausing or terminating prematurely. However, when a single round of transcription takes place on such templates in a test tube, RNA synthesis is frequently interrupted by pauses and premature terminations, which reduce the rate at which transcription takes place and frequently shorten the length of the mRNA molecules produced. Most pauses and premature terminations occur when RNA polymerase temporarily backtracks (i.e., backs up) for one or two nucleotides along the DNA. Experimental findings have demonstrated that most transcriptional delays and premature terminations disappear if several RNA polymerases are simultaneously transcribing the

DNA molecule. Propose an explanation for faster transcription and longer mRNA when the template DNA is being transcribed by multiple RNA polymerases.

Solution:
When a single RNA polymerase is transcribing a template in a test tube, the backtracking often may lead to the RNA polymerase and the 3′ end of the RNA transcript losing contact with each other resulting in premature termination. If other RNA polymerases are present on the template, they may act cooperatively with the leading RNA polymerase that has backtracked by pushing the leading RNA polymerase forward, thus continuing transcription before termination can occur.

Section 10.5

36.   [Data Analysis Problem] Alternative splicing takes place in more than 90% of the human genes that encode proteins. Researchers have found that how a pre-mRNA is spliced is affected by the pre-mRNA's promoter sequence (D. Auboeuf et al. 2002. *Science* 298:416–419). In addition, factors that affect the rate of elongation of the RNA polymerase during transcription affect the type of splicing that takes place. These findings suggest that the process of transcription affects splicing. Propose one or more mechanisms that would explain how transcription might affect alternative splicing.

Solution:
Transcription and splicing may be coupled. In other words, splicing reactions may begin before the transcription of the entire mRNA has completed. Components of the transcriptional complex may help recruit proteins needed for splicing to the splice sites, or spliceosomal proteins could participate in the transcriptional complex. The promoter sequence of the gene helps determine which transcription factors bind to the promoter. The interaction of the transcription complex and its proteins with splicing associated proteins may affect how the spliceosomes are located at splice sites, thus affecting splicing. Also, the rate at which RNA polymerase II proceeds could also affect transcription. If transcription occurs at a rapid pace, transcription may be completed before splicing occurs. Posttranscriptional splicing could result in different splice sites being used, possibly due to different types of secondary structure in the completed mRNA molecule.

37.   [Data Analysis Problem] Duchenne muscular dystrophy (DMD) is an X-linked recessive genetic disease caused by mutations in the gene that encodes dystrophin, a large protein that plays an important role in the development of normal muscle fibers. The gene that encodes dystrophin is immense, spanning 2.5 million base pairs, and includes 79 exons and 78 introns. Many of the mutations that cause DMD produce premature stop codons, which bring protein synthesis to a halt, resulting in a greatly shortened and nonfunctional form of dystrophin. Some geneticists have proposed treating DMD patients by causing the spliceosome to skip the exon containing the stop codon. Exon skipping will produce a protein that is somewhat shortened (because an exon is skipped and some amino acids are missing) but may still result in a protein that has some function (A. Goyenvalle et al. 2004. *Science* 306:1796–1799). Propose a mechanism to bring about exon skipping for the treatment of DMD.

Solution:

The proper splicing of the introns depends upon the sequences at the intron 5′ splice site, the branch site located upstream of the 3′ splice site, and the 3′ splice site. To ensure that the exon is removed by splicing, splice sites of the introns upstream and downstream of the exon need to be affected. One strategy would be to design antisense RNA molecules that can bind to the splice sites or branching points of particular introns. The target would be to block proper splicing of the introns surrounding the exon, but to allow a splicing event that would remove the exon containing the mutation along with its flanking introns. Antisense RNA molecules could be used to block splicing or block the binding of the necessary snRNPs. The use of antisense RNAs complementary to the 3′ splice site and/or the branch point of the intron upstream of the exon should block the 3′ splice from being cleaved, while an antisense RNA complementary to the 5′ splice site of the intron downstream of the exon should block the 5′ splice site from being cleaved. Ultimately, the upstream and downstream introns along with the exon sandwiched between will be removed, resulting in the exon being skipped in the production of the mature mRNA.

Chapter Eleven: From DNA to Proteins: Translation

COMPREHENSION QUESTIONS

Section 11.1

1. What is the one-gene, one-enzyme hypothesis?

   Solution:
   The one-gene, one-enzyme hypothesis proposed by Beadle and Tatum states that each gene encodes a single, separate protein. Now that we know more about the nature of enzymes and genes, it has been modified to the one-gene, one polypeptide hypothesis because many enzymes consist of multiple polypeptides. The original hypothesis helped establish a linear link between genes (DNA) and proteins.

2. What are isoaccepting tRNAs?

   Solution:
   Isoaccepting tRNAs are tRNA molecules that have different anticodon sequences but accept the same amino acids.

3. What is the significance of the fact that many synonymous codons differ only in the third nucleotide position?

   Solution:
   In synonymous codons that differ only at the third nucleotide position, the "wobble" and nonstandard base pairing with the anticodons will likely result in the correct amino acid being inserted in the protein.

4. Define the following terms as they apply to the genetic code:

   a. Reading frame

   Solution:
   The reading frame refers to how the nucleotides in a nucleic acid molecule are grouped into codons, with each codon containing three nucleotides. Any sequence of nucleotides has three potential reading frames that have completely different sets of codons.

   b. Overlapping code

   Solution:
   In an overlapping code, a single nucleotide is included in more than one codon. The result for a sequence of nucleotides is that more than one type of polypeptide can be encoded within that sequence.

c. Nonoverlapping code

Solution:
In a nonoverlapping code, a single nucleotide is part of only one codon, which results in the production of a single type of polypeptide from one polynucleotide sequence.

d. Initiation codon

Solution:
An initiation codon establishes the appropriate reading frame and specifies the first amino acid of the protein chain. Typically, the initiation codon is AUG; however, GUG and UUG can also serve as initiation codons.

e. Termination codon

Solution:
The termination codon signals the termination, or end, of translation and the end of the protein molecule. The three types of termination codons—UAA, UAG, and UGA—are also referred to as stop codons or nonsense codons. These codons do not code for amino acids.

f. Sense codon

Solution:
A sense codon is a group of three nucleotides that encode an amino acid. There are 61 sense codons that encode the 20 amino acids commonly found in proteins.

g. Nonsense codon

Solution:
A nonsense codon, or termination codon, signals the end of translation. These codons do not code for amino acids.

h. Universal code

Solution:
In a universal code, each codon specifies the same amino acid in all organisms. The genetic code is nearly universal but not completely. Most of the exceptions are in mitochondrial genes.

i. Nonuniversal codons

Solution:
Although most codons are universal (or nearly universal) in that they specify the same amino acids in almost all organisms, there are exceptions in which a codon has different meanings in different organisms.

5.   How is the reading frame of a nucleotide sequence set?

Solution:
The initiation codon on the mRNA sets the reading frame.

Section 11.2

6.   How are tRNAs linked to their corresponding amino acids?

Solution:
For each of the 20 different amino acids commonly found in proteins, a corresponding aminoacyl-tRNA synthetase covalently links the amino acid to the correct tRNA molecule.

7.   What role do the initiation factors play in protein synthesis?

Solution:
Initiation factors are proteins that are required for the initiation of translation. In bacteria, there are three initiation factors (IF1, IF2, and IF3). Each one has a different role. IF1 promotes the disassociation of the large and small ribosomal subunits. IF3 binds to the small ribosomal subunit and prevents it from associating with the large ribosomal subunit. IF2 is responsible for binding GTP and delivering the fMet-tRNA^{fMet} to the initiator codon on the mRNA. In eukaryotes, there are more initiation factors, but many have similar roles. Some of the eukaryotic initiation factors are necessary for recognition of the 5′ cap on the mRNA. Others possess a RNA helicase activity, which is necessary to resolve secondary structures.

8.   What events bring about the termination of translation?

Solution:
The process of termination begins when a ribosome encounters a termination codon. Because the termination codon would be located at the "A" site, no corresponding tRNA will enter the ribosome. This allows for the release factors (RF-1, RF-2, and RF-3) to bind the ribosome. RF-1 recognizes and interacts with the stop codons UAA and UAG, whereas RF-2 can interact with UAA and UGA. A RF-3-GTP complex binds to the ribosome. Termination of protein synthesis is complete when the polypeptide chain is cleaved from the tRNA located at the "P" site. During this process, the GTP is hydrolyzed to GDP.

9.   Compare and contrast the process of protein synthesis in bacterial and eukaryotic cells, giving similarities and differences in the process of translation in these two types of cells.

Solution:
Bacterial and eukaryotic cells share several similarities as well as have several differences in protein synthesis. Initially, bacteria and eukaryotes share the universal genetic code. However, the initiation codon, AUG, in eukaryotic cells codes for methionine, whereas in bacteria the AUG codon codes for *N*-Formylmethionine. In eukaryotes, transcription takes

place within the nucleus, whereas most translation takes place in the cytoplasm (although some translation does take place within the nucleus). Therefore, transcription and translation in eukaryotes are kept temporally and spatially separate. However, in bacterial cells transcription and translation occur nearly simultaneously.

Stability of mRNA in eukaryotic cells and bacterial cells is also different. Bacterial mRNA is typically short-lived, lasting only a few minutes. Eukaryotic mRNA may last hours or even days. Charging of the tRNAs with amino acids is essentially the same in both bacteria and eukaryotes. The ribosomes of bacteria and eukaryotes are different as well. Bacteria and eukaryotes have large and small ribosomal subunits, but they differ in size and composition. The bacterial large ribosomal consists of two ribosomal RNAs, whereas the eukaryotic large ribosomal subunit consists of three.

During translation initiation, the bacterial small ribosomal subunit recognizes the Shine-Dalgarno consensus sequence in the 5′ UTR of the mRNA and to regions of the 16S rRNA. In most eukaryotic mRNAs, the small subunit binds the 5′ cap of the mRNA and scans downstream until it encounters the first AUG codon. Finally, elongation and termination in bacterial and eukaryotic cells are functionally similar, although different elongation and termination factors are used.

Section 11.3

10.   What are some types of posttranslational modification of proteins?

Solution:
Several different modifications can occur to a protein following translation. Frequently, the amino terminal methionine may be removed. Sometimes, in bacteria only, the formyl group is cleaved from the *N*-formylmethionine, leaving a methionine at the amino terminal. More extensive modification occurs in some proteins that are originally synthesized as precursor proteins. These precursor proteins are cleaved and trimmed by protease enzymes to produce a functional protein. Glycoproteins are produced by the attachment of carbohydrates to newly synthesized proteins. Molecular chaperones are needed by many proteins to ensure that the proteins are folded correctly. Secreted proteins that are targeted for the membrane or other cellular locations frequently have 15 to 30 amino acids, called the signal sequence, removed from the amino terminal. Finally, acetylation of amino acids in the amino terminal of some eukaryotic proteins also occurs.

11.   Explain how some antibiotics work by affecting the process of protein synthesis.

Solution:
A number of antibiotics bind the ribosome and inhibit protein synthesis at different steps in translation. Some antibiotics, such as streptomycin, bind to the small subunit and inhibit translation initiation. Other antibiotics, such as chloramphenicol, bind to the large subunit and block the elongation of the peptide by preventing peptide-bond formation.

APPLICATION QUESTIONS AND PROBLEMS

Section 11.1

*12.   Assume that the number of different types of bases in RNA is four. What would be the minimum codon size (number of nucleotides) required to specify all amino acids if the number of different types of amino acids in proteins were: (a) 2, (b) 8, (c) 17, (d) 45, (e) 75?

Solution:
a.   1
b.   2
c.   3
d.   3
e.   4

*13.   How many codons would be possible in a triplet code if only three bases (A, C, and U) were used?

Solution:
To calculate the number of possible codons of a triplet code if only three bases are used, the following equation can be used: 3^n, where n is the number of nucleotides within the codon. So, the number of possible codons is equal to 3^3, or 27 possible codons.

14.   Using the genetic code presented in **Figure 11.5**, indicate which amino acid is encoded by each of the following mRNA codons.

Solution:
a.   5′–CCC –3′
         Pro
b.   5′–UUG–3′
         Leu
c.   5′–CUG–3′
         Leu
d.   5′–AGA–5′
         Arg
e.   5′–UAA–3′
         No amino acid, stop codon

*15.   Referring to the genetic code presented in **Figure 11.5**, give the amino acids specified by the following bacterial mRNA sequences, and indicate the amino and carboxyl ends of the polypeptide produced. Hint: Remember that AUG is the initiation codon.

Solution:

   a. 5′–<u>AUG</u> UUU AAA UUU AAA UUU UGA–3′

     Amino   fMet–Phe–Lys–Phe–Lys–Phe Carboxyl

   b. 5′–AGGGAAAUCAG <u>AUG</u> UAU AUA UAU AUA UGA–3′

     Amino   fMet–Tyr–Ile–Tyr–Ile carboxyl

   c. 5′–UUUGGAUUGAGUGAAACG <u>AUG</u> GAU GAA AGA UUU CUC GCU UGA–3

     Amino   fMet–Asp–Glu–Arg–Phe–Leu–Ala Carboxyl

   d. 5′–GUACUAAGGAGGUUGU <u>AUG</u> GGU UAG GGG ACA UCA UUU UGA–3′

     Amino   fMet–Gly Carboxyl

16. A nontemplate strand on bacterial DNA has the following base sequence. What amino acid sequence will be encoded by this sequence?
      5′–ATGATACTAAGGCCC–3′

Solution:
To determine the amino acid sequence, we need to know the mRNA sequence and the codons present. The nontemplate strand of the DNA has the same sequence as the mRNA, except that thymine-containing nucleotides are substituted for the uracil-containing nucleotides. So the mRNA sequence would be as follows:
      5′–AUGAUACUAAGGCCC–3′.

Assuming that the AUG indicates a start codon, then the amino acid sequence would be, starting from the amino end of the peptide and ending with the carboxyl end: fMet-Ile–Leu–Arg–Pro.

*17. The following amino acid sequence is found in a tripeptide: Met–Trp–His. Give all possible nucleotide sequences on the mRNA, on the template strand of DNA, and on the nontemplate strand of DNA that can encode this tripeptide.

Solution:
The amino acid His has two potential codons, whereas the amino acids Met and Trp each have only one potential codon. Therefore, there are two different mRNA nucleotide sequences that could encode for the tripeptide. Once the potential mRNA nucleotide sequences have been determined, the template and nontemplate DNA strands can be derived from these potential mRNA sequences.

(1) 5′–AUGUGGCAU–3′
    DNA template: 3′–TACACCGTA–5′
    DNA nontemplate: 5′–ATGUGGCAT–3′
(2) 5′–ATGUGGCAC–3′
    DNA template: 3′–TACACCGTG–5′
    DNA nontemplate: 5′–ATGTGGCAC–3′

18. How many different mRNA sequences can encode a polypeptide chain with the amino acid sequence Met–Leu–Arg? (Be sure to include the stop codon.)

Solution:
Leucine and arginine each have six different potential codons. There are also three potential stop codons. As for methionine, only one codon, AUG, is typically found as the initiation codon. (However, UUG and GUG have been shown to serve as start codons on occasion. For this problem, we will ignore these rare cases.) Therefore, the number of potential sequences is the product of the number of different potential codons for this tripeptide, which gives us a total of $(1 \times 6 \times 6 \times 3) = 108$ different mRNA sequences that can code for the tripeptide Met–Leu–Arg.

19. The following anticodons are found in a series of tRNAs. Refer to the genetic code in **Figure 11.5** and give the amino acid carried by each of these tRNAs.

Solution:
a. 5′–GUA–3′
    Tyr
b. 5′–AUU–3′
    Asn
c. 5′–GGU–3′
    Thr
d. 5′–CCU–3′
    Arg

20. Which of the following amino acid changes could result from a mutation that changed a single base? For each change that could result from the alteration of a single base, determine which position of the codon (first, second, or third nucleotide) in the mRNA must be altered for the change to result.

a. Leu → Gln

    Solution:
    Of the six codons that encode for Leu, only two could be mutated by the alteration of a single base to produce the codons for Gln:
        CUA (Leu)—Change the second position to A to produce CAA (Gln).
        CUG (Leu)—Change the second position to A to produce CAG (Gln).

b. Phe → Ser

Solution:
Both Phe codons (UUU and UUC) could be mutated at the second position to produce Ser codons:

UUU (Phe)—Change the second position to C to produce UCU (Ser).
UUC (Phe)—Change the second position to C to produce UCC (Ser).

c. Phe → Ile

Solution:
Both Phe codons (UUU and UUC) could be mutated at the first position to produce Ile codons:

UUU (Phe)—Change the first position to A to produce AUU (Ile).
UUC (Phe)—Change the first position to A to produce AUC (Ile).

d. Pro → Ala

Solution:
All four codons for Pro can be mutated at the first position to produce Ala codons:

CCU (Pro)—Change the first position to G to produce GCU (Ala).
CCC (Pro)—Change the first position to G to produce GCC (Ala).
CCA (Pro)—Change the first position to G to produce GCA (Ala).
CCG (Pro)—Change the first position to G to produce GCG (Ala).

e. Asn → Lys

Solution:
Both codons for Asn can be mutated at a single position to produce Lys codons:

AAU (Asn)—Change the third position to A to produce AAA (Lys).
AAU (Asn)—Change the third position to G to produce AAG (Lys).
AAC (Asn)—Change the third position to A to produce AAA (Lys).
AAC (Asn)—Change the third position to G to produce AAG (Lys).

f. Ile → Asn

Solution:
Only two of the three Ile codons can be mutated at a single position to produce Asn codons:

AUU (Ile)—Change the second position to A to produce AAU (Asn).
AUC (Ile)—Change the second position to A to produce AAC (Asn).

Section 11.2

*21. Arrange the following components of translation in the approximate order in which they would appear or be used in prokaryotic protein synthesis:

Solution:
The components are in order according to when they are used or play a key role in translation. The potential exception is initiation factor 3. Initiation factor 3 could possibly be listed first because it is necessary to prevent the 30S ribosome from associating with the 50S ribosome. It binds to the 30S subunit prior to the formation of the 30S initiation

complex. However, during translation events the release of initiation factor 3 allows the 70S initiation complex to form, a key step in translation.

Initiation factor 3
fMet-tRNA^{fMet}
30S initiation complex
70S initiation complex
Elongation factor Tu
Elongation factor G
Release factor 1

22.   The following diagram illustrates a step in the process of translation. Identify the following elements on the diagram.

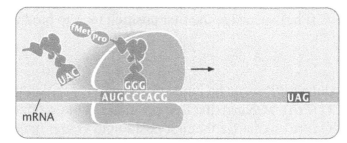

a. 5′ and 3′ ends of the mRNA
b. A, P, and E sites
c. Start codon
d. Stop codon
e. Amino and carboxyl ends of the newly synthesized polypeptide chain
f. Approximate location of the next peptide bond that will be formed
g. Place on the ribosome where release factor 1 will bind

Solution:

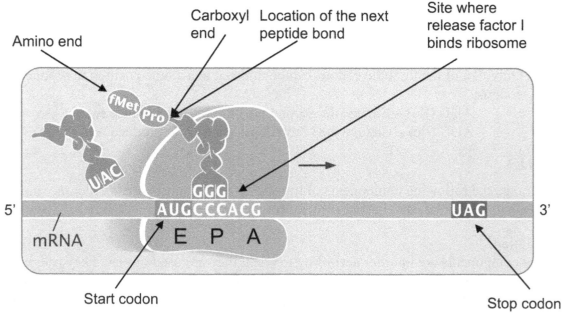

23.   Refer to the diagram in Problem 22 to answer the following questions.

   a.   What will be the anticodon of the next tRNA added to the A site of the ribosome?

   Solution:
   The anticodon 5′–CGU–3′ is complementary to the codon 5′–ACG–3′, which is located at the A site of the ribosome.

   b.   What will be the next amino acid added to the growing polypeptide chain?

   Solution:
   The codon 5′–ACG–3′ encodes the amino acid threonine.

*24.   A synthetic mRNA added to a cell-free protein-synthesizing system produces a peptide with the following amino acid sequence: Met–Pro–Ile–Ser–Ala. What would be the effect on translation if the following components were omitted from the cell-free protein- synthesizing system? What, if any, type of protein would be produced? Explain your reasoning.

   a.   Initiation factor 3

   Solution:
   The lack of IF-1 would decrease the amount of protein synthesized. IF-1 promotes the disassociation of the large and small ribosomal subunits. The initiation of translation requires a free small subunit. The absence of IF-1 would reduce the rate of initiation because more of the small ribosomal subunits would remain bound to the large ribosomal subunits.

   b.   Initiation factor 2

   Solution:
   No translation would occur. IF-2 is necessary for translation initiation. The lack of IF-2 would prevent fMet-tRNA^{fMet} from being delivered to the small ribosomal subunit, thus blocking translation.

   c.   Elongation factor Tu

   Solution:
   Although translation would be initiated by the delivery of the methionine to the ribosome-mRNA complex, no other amino acids would be delivered to the ribosome. EF-Tu, which binds to GTP and the charged tRNA, is necessary for elongation. This three-part complex enters the A site of the ribosome. If EF-Tu is not present, the charged tRNA will not enter the A site, thus stopping translation.

   d.   Elongation factor G

   Solution:
   EF-G is necessary for the translocation (movement) of the ribosome along the mRNA in a 5′ to 3′ direction. When a peptide bond has formed between the Met and Pro, the

lack of EF-G would prevent the movement of the ribosome along the mRNA, and so no new codons would be read. The formation of the dipeptide Met-Pro does not require EF-G.

e. Release factors RF-1, RF-2, and RF-3

Solution:
The release factors RF-1 and RF-2 recognize the stop codons and bind to the ribosome at the A site. They then interact with RF-3 to promote the cleavage of the peptide from the tRNA at the P site. The absence of the release factors would prevent the termination of translation at the stop codon.

f. ATP

Solution:
ATP is required for tRNAs to be charged with amino acids by aminoacyl-tRNA synthetases. Without ATP, the charging would not take place, and no amino acids will be available for protein synthesis.

g. GTP

Solution:
GTP is required for initiation, elongation, and termination of translation. If GTP is absent, protein synthesis will not take place.

25.   For each of the sequences in the following table, place a check mark in the appropriate space to indicate the process *most immediately* affected by deleting the sequence. Choose only one process for each sequence (i.e., one check mark per sequence).

Solution:

	Process most immediately affected by deletion			
Sequence deleted	**Replication**	**Transcription**	**RNA processing**	**Translation**
a. *ori* site The *ori* site or origin of replication is necessary for the initiation of replication.	✓	_____	_____	_____
b. 3′ splice-site consensus The 3′ splice site is necessary for proper excision of the intron. Therefore, RNA processing events will be affected.	_____	_____	✓	_____

c. Poly(A) tail
The poly(A) tail is involved in mRNA stability. If the tail is missing, then the mRNA will be degraded more rapidly, thus affecting translation.   ✓

d. Terminator
The terminator is necessary for transcription termination. Deletion of the terminator will result in the production of an abnormally long RNA transcript.   ✓

e. Start codon
The start codon is necessary for translation initiation.   ✓

f. –10 consensus
In bacteria, the –10 sequence is an important component of the promoter. Deletion of the –10 sequence will prevent transcription initiation from occurring.   ✓

g. Shine-Dalgarno
The Shine-Dalgarno sequence or ribosome binding site is bound by the 30S subunit during the initiation of translation. If the sequence is deleted, then the ribosome will not bind to the mRNA molecule and translation will not occur.   ✓

26.   Give the amino acid sequence of the protein encoded by the mRNA in **Figure 11.13**.

Solution:
Met-Pro-Thr-Thr-Ala-Ser-Val-Pro-Leu-Arg

CHALLENGE QUESTIONS

Section 11.1

27.   The redundancy of the genetic code means that some amino acids are specified by more than one codon. For example, the amino acid leucine is encoded by six different codons. Within a genome, synonymous codons are not present in equal numbers; some

synonymous codons appear much more frequently than others, and the preferred codons differ among different species. For example, in one species, the codon UUA might be used most often to encode leucine, whereas, in another species, the codon CUU might be used most often. Speculate on a reason for this bias in codon usage and why the preferred codons are not the same in all organisms.

Solution:
Synonymous codon usage patterns may depend on a variety of factors. Two potential factors that could affect usage patterns are the GC content of the organism and the relative amounts of isoaccepting tRNA molecules. The GC content of an organism reflects the relative proportions of nucleotides found in the DNA. In a given organism, the bias for particular synonymous codons may reflect the overall GC content of that organism. For example, in organisms that have a high GC content, you might expect to find that the synonymous codon usage pattern reflects this bias, resulting in a preference for codons with more Gs and Cs. Therefore, in two organisms that differ in GC content, the synonymous codon usage bias should reflect their differences in base composition.

Isoaccepting tRNAs are those that carry the same amino acid but have different anticodons. These isoaccepting tRNAs act at synonymous codons. For a given organism, the more prevalent synonymous codons may depend on the frequency of its tRNA and complementary anticodon. In different organisms, the concentrations of the various isoaccepting tRNAs will vary leading to different usage patterns of synonymous codons in these organisms.

Section 11.2

*28. Several experiments were conducted to obtain information about how the eukaryotic ribosome recognizes the AUG start codon. In one experiment, the gene that encodes methionine initiator tRNA ($tRNA_i^{Met}$) was located and changed. The nucleotides that specify the anticodon on $tRNA_i^{Met}$ were mutated so that the anticodon in the tRNA was 5′–CCA–3′ instead of 5′–CAU–3′. When this mutated gene was placed into a eukaryotic cell, protein synthesis took place but the proteins produced were abnormal. Some of the proteins produced contained extra amino acids, and others contained fewer amino acids than normal.

   a. What do these results indicate about how the ribosome recognizes the starting point for translation in eukaryotic cells? Explain your reasoning.

   Solution:
   The results suggest that, to initiate translation, the ribosome scans the mRNA to find the appropriate start sequence.

b. If the same experiment had been conducted on bacterial cells, what results would you expect?

Solution:
The initiation of translation in bacteria requires the 16S RNA of the small ribosomal subunit to interact with the Shine-Dalgarno sequence. This interaction serves to line up the ribosome over the start codon. If the anticodon has been changed such that the start codon cannot be recognized, then protein synthesis is not likely to take place.

c. Explain why some proteins contained extra amino acids while others contained fewer amino acids than normal.

Solution:
If the first 5′–UGG–3′ codon occurs prior to the normal 5′–AUG–3′ codon, then a protein containing extra amino acids could be produced

Chapter Twelve: Control of Gene Expression

COMPREHENSION QUESTIONS

Section 12.1

1. Why is gene regulation important for bacterial cells?

   Solution:
   Gene regulation allows for biochemical and internal flexibility while maintaining energy efficiency by the bacterial cells.

2. Name six different levels at which gene expression might be controlled.

   Solution:
   (1) Alteration or modification of the gene structure at the DNA level
   (2) Transcriptional regulation
   (3) Regulation at the level of mRNA processing
   (4) Regulation of mRNA stability
   (5) Regulation of translation
   (6) Regulation by post-translational modification of the synthesized protein

Section 12.2

3. Draw a picture illustrating the general structure of an operon and identify its parts.

   Solution:

4. What is the difference between positive and negative control? What is the difference between inducible and repressible operons?

   Solution:
   Positive transcriptional control requires an activator protein to stimulate transcription at the operon. In negative control, a repressor protein inhibits or turns off transcription at the operon.

   An inducible operon normally is not transcribed. It requires an inducer molecule to stimulate transcription either by inactivating a repressor protein in a negative inducible operon or by stimulating the activator protein in a positive inducible operon.

Transcription normally occurs in a repressible operon. In a repressible operon, transcription is turned off either by the repressor becoming active in a negative repressible operon or by the activator becoming inactive in a positive repressible operon.

5.   Briefly describe the *lac* operon and how it controls the metabolism of lactose.

Solution:
The *lac* operon consists of three structural genes—the *lacZ* gene, the *lacY* gene, and the *lacA* gene, which encode β-galactosidase, permease, and thiogalactoside transacetylase, respectively. All three genes share a promoter and operator region. Upstream from the lactose operon is the *lacI* gene that encodes the *lac* operon repressor, which binds at the operator region and inhibits transcription of the *lac* operon by preventing RNA polymerase from successfully initiating transcription.

When lactose is present in the cell, the enzyme β-galactosidase converts some of it into allolactose, which binds to the *lac* repressor, altering its shape and reducing the repressor's affinity for the operator. Because the allolactose-bound repressor does not bind to the operator, RNA polymerase can initiate transcription of the *lac* structural genes from the lac promoter.

6.   What is catabolite repression? How does it allow a bacterial cell to use glucose in preference to other sugars?

Solution:
In catabolite repression, the presence of glucose inhibits or represses the transcription of genes involved in the metabolism of other sugars. Because the gene expression necessary for utilizing other sugars is turned off, only enzymes involved in the metabolism of glucose will be synthesized. Operons that exhibit catabolite repression are under the positive control of catabolic activator protein (CAP). For CAP to be active, it must form a complex with cAMP. Glucose affects the level of cAMP. The levels of glucose and cAMP are inversely proportional—as glucose levels increase, the level of cAMP decreases. Thus, CAP is not activated.

Section 12.3

7.   What changes take place in chromatin structure and what role do these changes play in eukaryotic gene regulation?

Solution:
Changes in chromatin structure can result in repression or stimulation of gene expression. The acetylation of histone proteins increases transcription. The reverse reaction by deacetylases restores repression. Chromatin-remodeling complexes bind directly to the DNA, altering chromatin structure without acetylating histone proteins and allowing transcription to be initiated by making the promoters accessible to transcriptional factors. The methylation of DNA sequences represses transcription. The demethylation of DNA sequences often increases transcription.

8.   What is the histone code?

Solution:
The histone code refers to modifications to the tails of histone proteins. These modifications include the addition or removal of phosphate groups, acetyl groups, or methyl groups to the tails. Information imparted by these modifications affects how genes are expressed.

9.   Briefly explain how transcriptional activator and repressor proteins affect the level of transcription of eukaryotic genes.

Solution:
Transcriptional activator proteins stimulate transcription by binding DNA at specific base sequences such as an enhancer or regulatory promoter and attracting or stabilizing the basal transcriptional factor apparatus. Repressor proteins bind to silencer sequences or promoter regulator sequences. These proteins may inhibit transcription by blocking access to the enhancer sequence by the activator protein, preventing the activator from interacting with the basal transcription apparatus, or preventing the basal transcription factor from being assembled.

10.   What is an enhancer? How does it affect transcription of distant genes?

Solution:
An enhancer is a DNA sequence that, when bound to transcriptional activator proteins, can affect the transcription of a distant gene. Transcription at a distant gene is affected when the DNA sequence between the gene's promoter and the enhancer loops out, bringing the promoter and the enhancer close together and allowing the transcriptional activator proteins to directly interact with the basal transcription apparatus at the promoter, which stimulates transcription.

11.   What role does mRNA stability play in gene regulation? What controls mRNA stability in eukaryotic cells?

Solution:
The total amount of protein synthesized depends on the amount of mRNA available for translation. The amount of available mRNA depends on the rates of mRNA synthesis and degradation. Less-stable mRNAs degrade faster than stable mRNAs, and so fewer copies of the mRNA are available as templates for translation.

The 5′ cap, 3′ poly(A) tail, the 5′ UTR, 3′ UTR, and the coding region in the mRNA molecule affect its stability. Poly(A)-binding proteins bind at the 3′ poly(A) tail. These proteins contribute to the stability of the tail and protect the 5′ cap through direct interaction. When a critical number of adenine nucleotides have been removed from the tail, the protection is lost and the 5′ cap is removed. The removal of the 5′ cap enables 5′-to-3′ nucleases to degrade the mRNA.

12. Briefly list some of the ways in which siRNAs and miRNAs regulate genes.

Solution:
Through Slicer activity, which cleaves mRNA sequences; through the binding of miRNAs to complementary regions in mRNA, which prevents translation; and through transcriptional silencing, in which siRNAs play a role in altering chromatin structure.

13. How does bacterial gene regulation differ from eukaryotic gene regulation? How are they similar?

Solution:
Much of gene regulation in bacteria occurs at the level of transcription, whereas gene regulation in eukaryotes often takes place at multiple levels. Modification of chromatin structure plays an important role in regulating eukaryotic transcription; chromatin structure is absent in bacteria. Bacterial genes are often organized in operons and coordinately expressed. In contrast, most eukaryotic genes have their own promoters. Eukaryotic transcription is controlled by a machinery more complex than that of bacteria; this complex machinery includes numerous transcription factors and transcriptional activators. RNA processing plays a larger role in eukaryotic gene regulation. Small RNA molecules (siRNAs and miRNAs) play an important role in eukaryotic gene regulation but are absent from most bacteria.

Section 12.4

14. What are epigenetic effects? How do they differ from other genetic traits?

Solution:
Epigenetic effects are differences in the expression of genes that are passed on to other cells and sometimes to other generations. They are due to alterations in DNA and chromatin structure but without changes of the DNA nucleotide sequence.

15. What types of changes are thought to be responsible for epigenetic traits?

Solution:
Changes in chromatin structure, brought about by DNA methylation, the alteration of histone proteins, and the reposition of nucleosomes.

16. How are patterns of DNA methylation maintained across cell divisions?

Solution:
Methyltransferase enzymes recognize the hemimethylated state of CpG dinucleotides following replication and add methyl groups to the unmethylated cytosines, resulting in two new DNA molecules that are fully methylated.

APPLICATION QUESTIONS AND PROBLEMS

Section 12.2

*17. For each of the following types of transcriptional control, indicate whether the protein produced by the regulator gene will be synthesized initially as an active repressor, inactive repressor, active activator, or inactive activator.

Solution:

a. Negative control in a repressible operon
   Inactive repressor

b. Positive control in a repressible operon
   Active activator

c. Negative control in an inducible operon
   Active repressor

d. Positive control in an inducible operon
   Inactive activator

*18. A mutation at the operator prevents the regulator protein from binding. What effect will this mutation have in the following types of operons?

Solution:

a. Regulator protein is a repressor in a repressible operon.
   The operon would never be turned off, and transcription will take place all the time.
b. Regulator protein is a repressor in an inducible operon.
   The result will be constitutive expression, and the transcription will take place all the time.

19. The *blob* operon produces enzymes that convert compound A into compound B. The operon is controlled by a regulatory gene *S*. Normally, the enzymes are synthesized only in the absence of compound B. If gene *S* is mutated, the enzymes are synthesized in the presence *and* in the absence of compound B. Does gene *S* produce a regulatory protein that exhibits positive or negative control? Is this operon inducible or repressible?

Solution:
Because the *blob* operon is transcriptionally inactive in the presence of B, gene *S* most likely codes for a regulatory protein that exhibits negative control. The data suggest that the *blob* operon is repressible because it is inactive in the presence of compound B, but active when compound B is absent.

*20. A mutation prevents the catabolite activator protein (CAP) from binding to the promoter in the *lac* operon. What will the effect of this mutation be on the transcription of the operon?

Solution:
RNA polymerase will bind to the *lac* promoter poorly, significantly decreasing the transcription of the *lac* structural genes.

21. Under which of the following conditions would a *lac* operon produce the greatest amount of β-galactosidase? The least? Explain your reasoning.

Solution:

	Lactose present	Glucose present
Condition 1	Yes	No
Condition 2	No	Yes
Condition 3	Yes	Yes
Condition 4	No	No

Condition 1 will result in the production of the maximum amount of β-galactosidase. For maximum transcription, the presence of lactose and the absence of glucose are required. Lactose (or allolactose) binds to the *lac* repressor reducing the affinity of the *lac* repressor to the operator. This decreased affinity results in the promoter being accessible to RNA polymerase. The lack of glucose allows for increased synthesis of cAMP, which can complex with CAP. The formation of CAP-cAMP complexes improves the efficiency of RNA polymerase binding to the promoter, which results in higher levels of transcription from the *lac* operon.

Condition 2 will result in the production of the least amount of β-galactosidase. With no lactose present, the *lac* repressor is active and binds to the operator, inhibiting transcription. The presence of glucose results in a decrease of cAMP levels. A CAP-cAMP complex does not form, and RNA polymerase will not be stimulated to transcribe the lac operon.

22. A mutant strain of *E. coli* produces β-galactosidase in the presence *and* in the absence of lactose. Where in the operon might the mutation in this strain be located?

Solution:
Within the operon, the operator region is the most probable location of the mutation. If the mutation prevents the *lac* repressor protein from binding to the operator, then transcription of the *lac* structural genes will not be inhibited. Expression will be constitutive.

23. Examine **Figure 12.7.** What would be the effect of a drug that altered the structure of allolactose so that it was unable to bind to the regulator protein?

Solution:
Allolactose is produced when lactose is present; allolactose normally binds to the repressor protein and makes it inactive, allowing transcription to occur when lactose is present. If a drug altered the structure of allolactose, it would not bind to the repressor and the repressor would continue to bind to the operator, keeping transcription off. The result

would be that transcription was repressed even in the presence of lactose; thus, no β-galactosidase or permease would be produced.

*24. For *E. coli* strains with the *lac* genotypes given below, use a plus sign (+) to indicate the synthesis of β-galactosidase and permease and a minus sign (–) to indicate no synthesis of the proteins.

Solution:
In determining if expression of the β-galactosidase and the permease gene will occur, you should consider several factors. The presence of *lacZ⁺* and *lacY⁺* on the same DNA molecule as a functional promoter (*lacP⁺*) is required because the promoter is a cis acting regulatory element. However the *lacI⁺* gene product or *lac* repressor is trans acting and does not have to be located on the same DNA molecule as β-galactosidase and permease genes to inhibit expression. For the repressor to function, it does require that the cis acting *lac* operator be on the same DNA molecule as the functional β-galactosidase and permease genes. Finally, the dominant *lacIˢ* gene product is also trans acting and can inhibit transcription at any functional *lac* operator region.

Genotype of strain	Lactose absent		Lactose present	
	β-Galactosidase	Permease	β-Galactosidase	Permease
lacI⁺ lacP⁺ lacO⁺ lacZ⁺ lacY⁺	–	–	+	+
lacI⁻ lacP⁺ lacO⁺ lacZ⁺ lacY⁺	+	+	+	+
lacI⁺ lacP⁺ lacOᶜ lacZ⁺ lacY⁺	+	+	+	+
lacI⁻ lacP⁺ lacO⁺ lacZ⁺ lacY⁻	+	–	+	–
lacI⁻ lacP⁻ lacO⁺ lacZ⁺ lacY⁺	–	–	–	–
lacI⁺ lacP⁺ lacO⁺ lacZ⁻ lacY⁺ / *lacI⁻ lacP⁺ lacO⁺ lacZ⁺ lacY⁻*	–	–	+	+
lacI⁻ lacP⁺ lacOᶜlacZ⁺ lacY⁺ / *lacI⁺ lacP⁺ lacO⁺ lacZ⁻ lacY⁻*	+	+	+	+
lacI⁻ lacP⁺ lacO⁺ lacZ⁺ lacY⁻ / *lacI⁺ lacP⁻ lacO⁺ lacZ⁻ lacY⁺*	–	–	+	–
lacI⁺ lacP⁻ lacOᶜlacZ⁻ lacY⁺ / *lacI⁻ lacP⁺ lacO⁺ lacZ⁺ lacY⁻*	–	–	+	–
lacI⁺ lacP⁺ lacO⁺ lacZ⁺ lacY⁺ / *lacI⁺ lacP⁺ lacO⁺ lacZ⁺ lacY⁺*	–	–	+	+
lacIˢlacP⁺ lacO⁺ lacZ⁺ lacY⁻ / *lacI⁺ lacP⁺ lacO⁺ lacZ⁻ lacY⁺*	–	–	–	–
lacIˢlacP⁻ lacO⁺ lacZ⁻ lacY⁺/ *lacI⁺ lacP⁺ lacO⁺ lacZ⁺ lacY⁺*	–	–	–	–

25. Give all possible genotypes of a *lac* operon that produces β-galactosidase and permease under the following conditions. Do not give partial-diploid genotypes.

Solution:

	Lactose absent		**Lactose present**		
	β-Galactosidase	**Permease**	**β-Galactosidase**	**Permease**	**Genotype**
a.	–	–	+	+	*lacI⁺ lacP⁺ lacO⁺ lacZ⁺ lacY⁺*
b.	–	–	–	+	*lacI⁺ lacP⁺ lacO⁺lacZ⁻ lacY⁺*
c.	–	–	+	–	*lacI⁺ lacP⁺ lacO⁺ lacZ⁺ lacY⁻*
d.	+	+	+	+	*lacI⁻ lacP⁺ lacO⁺ lacZ⁺ lacY⁺*
					or
e.	–	–	–	–	*lacI⁺ lacP⁺ lacOᶜlacZ⁺lacY⁺*
					lacIˢ lacP⁺lacO⁺ lacZ⁺ lacY⁺
					or
f.	+	–	+	–	*lacI⁺ lacP⁻ lacO⁺ lacZ⁺ lacY⁺*
					lacI⁻ lacP⁺ lacO⁺ lacZ⁺ lacY⁻
					or
g.	–	+	–	+	*lacI⁺ lacP⁺ lacOᶜ lacZ⁺ lacY⁻*
					lacI⁻ lacP⁺ lacO⁺ lacZ⁻ lacY⁺
					or
					lacI⁺ lacP⁺ lacOᶜ lacZ⁻ lacY⁺

*26. Explain why mutations in the *lacI* gene are trans in their effects, but mutations in the *lacO* gene are cis in their effects.

Solution:
The *lacI* gene encodes the *lac* repressor protein, which can diffuse within the cell and attach to any operator. It can therefore affect the expression of genes on the same or on a different molecule of DNA. The *lacO* gene encodes the operator. It affects the binding of RNA polymerase to the DNA and therefore affects the expression of genes only on the same molecule of DNA.

27. The *mmm* operon, which has sequences *A, B, C,* and *D* (which may be structural genes or regulatory sequences), encodes enzymes 1 and 2. Mutations in sequences *A, B, C,* and *D* have the following effects, where a plus sign (+) indicates that the enzyme is synthesized and a minus sign (–) indicates that the enzyme is not synthesized.

	Mmm absent		**Mmm present**	
Mutation in sequence	**Enzyme 1**	**Enzyme 2**	**Enzyme 1**	**Enzyme 2**
No mutation	+	+	–	–
A	–	+	–	–
B	+	+	+	+
C	+	–	–	–
D	–	–	–	–

a. Is the *mmm* operon inducible or repressible?

Solution:
This is typical of a repressible operon.

b. Indicate which sequence (*A, B, C,* or *D*) is part of the following components of the operon:

Solution:
Regulator gene   _B_
When sequence B is mutated, gene expression is not repressed by the presence of mmm.
Promoter   _D_
When sequence D is mutated, no gene expression occurs either in the presence or absence of mmm.
Structural gene for enzyme 1   _A_
When sequence A is mutated, enzyme 1 is not produced.
Structural gene for enzyme 2   _C_
When sequence C is mutated, enzyme 2 is not produced.

28.   [Data Analysis Problem] Ellis Engelsberg and his coworkers examined the regulation of genes taking part in the metabolism of arabinose, a sugar (E. Engelsberg et al. 1965. *Journal of Bacteriology* 90:946–957). Four structural genes encode enzymes that help metabolize arabinose (genes *A, B, D,* and *E*). An additional gene *C* is linked to genes *A, B,* and *D*. These genes are in the order *D-A-B-C.* Gene *E* is distant from the other genes. Engelsberg and his colleagues isolated mutations at the *C* gene that affected the expression of structural genes *A, B, D,* and *E.* In one set of experiments, they created various genotypes at the *A* and *C* loci and determined whether arabinose isomerase (the enzyme encoded by gene *A*) was produced in the presence or absence of arabinose (the substrate of arabinose isomerase). Results from this experiment are shown in the following table, where a plus sign (+) indicates that the arabinose isomerase was synthesized and a minus sign (–) indicates that the enzyme was not synthesized.

Genotype	Arabinose absent	Arabinose present
1. $C^+ A^+$	–	+
2. $C^- A^+$	–	–
3. $C^- A^+/C^+ A^-$	–	+
4. $C^c A^-/C^- A^+$	+	+

a. On the basis of the results of these experiments, is the *C* gene an operator or a regulator gene? Explain your reasoning.

Solution:
The *C* gene is a regulator gene. The *C* gene is trans acting, thus it affects the expression of the *A* gene located on a different DNA molecule, which is typical of a

gene encoding a regulatory protein. If the *C* gene was an operator, it would be cis acting and only able to regulate the expression of the *A* gene found on the same DNA molecule, which is not the case as demonstrated from genotype 3.

b. Do these experiments suggest that the arabinose operon is negatively or positively controlled? Explain your reasoning.

Solution:
Data from these experiments suggest that the arabinose operon is positively controlled. From the above data, the *C* gene appears to be a regulator gene that is needed for the transcription of the *A* gene. For the *A* gene to be expressed, a functional *C* gene needs to be present within the cell. In the absence of a functional *C* gene and arabinose, the *A* gene is not expressed. Both results would be explained if the *C* gene encodes a regulator protein that is required to activate transcription of the *A* gene.

c. What type of mutation is C^c?

Solution:
The C^c mutation results in continuous activation of transcription from the *A* gene. In other words, C^c leads to constitutive expression of the *A* gene.

29. [Data Analysis Problem] In *E. coli,* three structural genes (*A, D,* and *E*) encode enzymes A, D, and E, respectively. Gene *O* is an operator. The genes are in the order *O-A-D-E* on the chromosome. These enzymes catalyze the biosynthesis of valine. Mutations were isolated at the *A, D, E,* and *O* genes to study the production of enzymes A, D, and E (T. Ramakrishnan and E. A. Adelberg. 1965. *Journal of Bacteriology* 89:654–660). Levels of the enzymes produced by partial-diploid *E. coli* with various combinations of mutations are shown in the following table.

Genotype	Amount of enzyme produced		
	E	D	A
1. $E^+\ D^+\ A^+\ O^+/$			
$E^+\ D^+\ A^+\ O^+$	2.40	2.00	3.50
2. $E^+\ D^+\ A^+\ O^-/$			
$E^+\ D^+\ A^+\ O^+$	35.80	38.60	46.80
3. $E^+\ D^-\ A^+\ O^-/$			
$E^+\ D^+\ A^-\ O^+$	1.80	1.00	47.00
4. $E^+\ D^+\ A^-\ O^-/$			
$E^+\ D^-\ A^+\ O^+$	35.30	38.00	1.70
5. $E^-\ D^+\ A^+\ O^-/$			
$E^+\ D^-\ A^+\ O^+$	2.38	38.00	46.70

a. Is the regulator protein that binds to the operator of this operon a repressor (negative control) or an activator (positive control)? Explain your reasoning.

Solution:
The regulator protein is a repressor. When the operator is defective or nonfunctioning (O^-), then the expression of enzymes encoded by the A, D, and E loci is significantly increased over the wild-type operator genotype. This suggests that the regulator protein cannot bind the O^- region and repress transcription.

b. Are genes A, D, and E all under the control of operator O? Explain your reasoning.

Solution:
Yes, all the genes are under control of the operator. When operator (O^+) and repressor are both functional, then the levels of expression are low. When the operator is nonfunctional, the enzyme levels generally increase.

c. Propose an explanation for the low level of enzyme E produced in genotype 3.

Solution:
The low level of enzyme E produced in genotype 3 is likely due to a polarity effect. Because they share the same transcriptional control regions, genes A, D, and E are transcribed together producing a single polycistronic mRNA molecule. Gene E is located downstream of gene D and is thus transcribed after gene D. In genotype 3, it is likely that the defect in gene D affects transcription elongation that occurs subsequent to the mutation.

Section 12.3

30. A geneticist is trying to determine how many genes are found in a 300,000-bp region of DNA. Analysis shows that four different areas within the 300,000-bp region have H3K4me3 modifications. What might their presence suggest about the number of genes located there?

Solution:
The histone 3 methylase H3K4me3 adds 3 methyl groups to the lysine 4 in the tail of histone 3. These modifications typically occur near the transcription start site of genes. If four H3K4me3 modifications have been identified, it suggests that at least four genes are present within this region of DNA.

31. In a line of human cells grown in culture, a geneticist isolates a temperature-sensitive mutation at a locus that encodes an acetyltransferase enzyme; at temperatures above 38°C, the mutant cells produce a nonfunctional form of the enzyme. What would be the most likely effect of this mutation when the cells are grown at 40°C?

Solution:
Acetyltransferase enzymes add acetyl groups to histone proteins preventing the proteins from forming the 30-nm chromatin fiber. Essentially, the chromatin structure is destabilized, which allows for transcription to occur. If the cells are raised to 40°C, then the acetyltransferase enzyme would not function and acetyl groups would not be added to the histone proteins that are the target of this enzyme. The result would be that the nucleosomes and the chromatin would remain stabilized and block transcriptional activation.

32. X31b is an experimental compound that is taken up by rapidly dividing cells. Research has shown that X31b stimulates the methylation of DNA. Some cancer researchers are interested in testing X31b as a possible drug for treating prostate cancer. Offer a possible explanation for why X31b might be an effective anticancer drug.

Solution:
Cancer cells are typically rapidly dividing cells. DNA methylation particularly in regions with many CpG sequences (CpG islands) is associated with transcriptional repression. If the X31b molecules can be uptaken by the rapidly dividing cancers cells and then stimulate methylation of DNA sequences in the cancer cells, transcriptional repression of genes in the cancer cells would be expected. The repression of transcription could affect the growth of the cancer cells and potentially cause a loss of viability of these cells.

33. What would be the effect of moving the insulator shown in **Figure 12.18** to a position between enhancer II and the promoter for gene *B*?

Solution:
Enhancers I and II would now stimulate gene *A*. Neither enhancer would stimulate gene *B*.

*34. An enhancer is surrounded by four genes (*A, B, C,* and *D*), as shown in the diagram below. An insulator lies between gene *C* and gene *D*. On the basis of the positions of the genes, the enhancer, and the insulator, the transcription of which genes is most likely to be stimulated by the enhancer? Explain your reasoning.

Gene A	Gene B	Enhancer	Gene C	Insulator	Gene D

Solution:
The action of an enhancer is blocked when the insulator is located between the enhancer and the promoter of the gene. It is likely that genes *A, B,* and *C* will be stimulated by the enhancer and that gene *D* will not be stimulated. Insulators block the stimulatory action of enhancers when they lie between the enhancer and the promoter of the gene. In the example from the figure, the insulator is only between gene *D* and the enhancer. The enhancer's effect on genes *A, B,* and *C* is not likely to be affected by the insulator, and these genes will be stimulated.

35.   Some eukaryotic mRNAs have an AU-rich element in the 3′ untranslated region. What would be the effect on gene expression if this element were mutated or deleted?

Solution:
The presence of AU-rich elements is associated with rapid degradation of the mRNA molecules that contain them through a RNA silencing mechanism. If the AU-element was deleted, then the miRNA would not be able to bind to the consensus sequence of the AU-rich element and the RISC degradation would not be initiated. It is likely that this mRNA molecule would be more stable resulting in increased gene expression of the protein coded for by the mRNA.

36.   A strain of *Arabidopsis thaliana* possesses a mutation in the *APETALA2* gene, in which much of the 3′ untranslated region of mRNA transcribed from the gene is deleted. What is the most likely effect of this mutation on the expression of the *APETALA2* gene?

Solution:
Translation from the *APETALA2* is inhibited by a miRNA that binds within the coding region of the mRNA. Thus, deleting much of the 3′ untranslated region of the *APETALA2* mRNA will likely not affect the translation regulation by the miRNA molecule. However, the 3′ untranslated region could potentially be needed for mRNA stability and binding of the ribosome to the mRNA molecule, so the deletion could result in a decrease in expression of the *APETALA2* gene.

Section 12.4

*37.   How do epigenetic traits differ from traditional genetic traits, such as the differences in color and shape of peas that Mendel studied?

Solution:
The phenotypic differences in traditional genetic traits such as the color and shape of peas that Mendel studied are due to differences in the DNA base sequences within the alleles. In epigenetics, the phenotypic differences are not due to changes in allele DNA base sequences, but are differences in the expression of genes that are passed on to other cells and sometimes to other generations.

*38.   A scientist does an experiment in which she removes the offspring of rats from their mother at birth and has her genetics students feed and rear the offspring. Assuming that the students do not lick and groom the baby rats like their mothers normally do, what long-term behavioral and epigenetic effects would you expect to see in the rats when they grow up?

Solution:
We would also expect that the adults would show increased fear and heightened hormonal response to stress. We would expect to see differences in DNA methylation and histone acetylation that altered expression of genes involved in response to stress.

39. [Data Analysis Problem] Pregnant female rats were exposed to a daily dose of 100 or 200 mg/kg of vinclozolin, a fungicide commonly used in the wine industry (M. D. Anway et al. 2005. *Science* 308:1466–1469). The F_1 offspring from the exposed female rats were interbred, producing F_2, F_3, and F_4 rats. None of the F_2, F_3, or F_4 rats were exposed to vinclozolin. Testes from the F_1–F_4 male rats were examined and compared with those of control rats descended from females that had not been exposed to vinclozolin. There were higher percentages of apoptotic cells (cells that underwent controlled cell death) in the testes of F_1–F_4 male descendants of females who were exposed to vinclozolin than in descendants of control females (graph a). Furthermore, sperm numbers (graph b) and motilities (graph c) were lower in the F_1–F_4 descendants of vinclozolin-exposed females than in those of control females. In addition, 8% of the F_1–F_4 males descended from vinclozolin-exposed females developed complete infertility, compared with 0% of the F_1–F_4 males descended from control females. Molecular analysis of the testes demonstrated that DNA methylation patterns differed between descendants of vinclozolin-exposed females and descendants of control females. Explain the transgenerational effects of vinclozolin on male fertility.

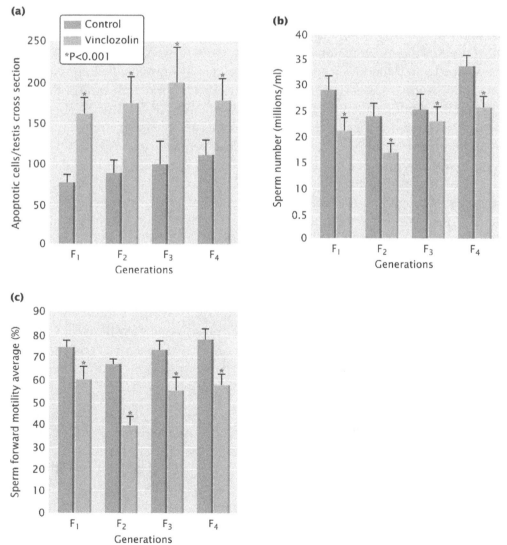

[Graphs after M. D. Anway et al. 2005. *Science* 308:1466–1469.]

Solution:
Because only the original pregnant females were exposed to vinclozolin, the effects on the sperm of F_2–F_4 mice cannot be explained by direct effects of vinclozolin on male fertility. Furthermore, because of the high frequency (90%) of the mice affected in F_2–F_4, it appears unlikely that the effects are due to mutations induced by vinclozolin. These transgenerational effects are most likely due to epigenetic changes. This is supported by the different DNA methylation patterns of the F_1–F_4 offspring of vinclozolin-exposed females. DNA methylation is known to affect chromatin structure and is responsible for some epigenetic effects.

CHALLENGE QUESTIONS

Section 12.3

40. [Data Analysis Problem] A yeast gene termed *SER3*, which has a role in serine biosynthesis, is repressed during growth in nutrient-rich medium, and so little transcription takes place and little SER3 enzyme is produced. In an investigation of the nature of the repression of the *SER3* gene, a region of DNA upstream of the *SER3* gene was found to be heavily transcribed when the *SER3* gene is repressed (J. A. Martens, L. Laprade, and F. Winston. 2004. *Nature* 429:571–574). Within this upstream region is a promoter that stimulates the transcription of an RNA molecule called *SRG1* RNA (for *SER3* regulatory gene 1). This RNA molecule has none of the sequences necessary for translation. Mutations in the promoter for *SRG1* result in the disappearance of *SRG1* RNA, and these mutations remove the repression of *SER3*. When RNA polymerase binds to the *SRG1* promoter, the polymerase has been found to travel downstream, transcribing the *SGR1* RNA, and to pass through and transcribe the promoter for *SER3*. This activity leads to the repression of *SER3*. Propose a possible explanation for how the transcription of *SGR1* might repress the transcription of *SER3*. (Hint: Remember that the *SGR1* RNA does not encode a protein.)

Solution:
Potentially, the transcription from the *SRG1* gene interferes with the transcription initiation of the *SER3* gene. Because part of the *SRG1* transcript overlaps with the *SER3* promoter, the transcriptional complex that is transcribing the *SRG1* gene may prevent transcriptional factors from interacting with the *SER3* promoter. Therefore, as long as the *SRG1* gene is being transcribed, the *SER3* transcription will not be initiated.

41. A common feature of many eukaryotic mRNAs is the presence of a rather long 3′ UTR, which often contains consensus sequences. Creatine kinase B (CK-B) is an enzyme important in cellular metabolism. Certain cells—termed U937D cells—have lots of CK-B mRNA, but no CK-B enzyme is present. In these cells, the 5′ end of the CK-B mRNA is bound to ribosomes, but the mRNA is apparently not translated. Something inhibits the translation of the CK-B mRNA in these cells.

Researchers introduced numerous short segments of RNA containing only 3′ UTR sequences into U937D cells. As a result, the U937D cells began to synthesize the CK-B enzyme, but the total amount of CK-B mRNA did not increase. The introduction of short segments of other RNA sequences did not stimulate the synthesis of CK-B; only the 3′ UTR sequences turned on the translation of the enzyme.

Based on these results, propose a mechanism for how CK-B translation is inhibited in U937D cells. Explain how the introduction of short segments of RNA containing the 3′ UTR sequences might remove the inhibition.

Solution:
From the above experimental data, translation of the CK-B protein is inhibited in the U937D cells—the CK-B mRNA is present and bound to the ribosome, but no protein is synthesized. A possible mechanism for the inhibition of translation could be the binding of translational repressors to the 3′ UTR region of the CK-B mRNA. The action of soluble proteins inhibiting translation seems to be suggested by the response of the U937 cells to the short RNA sequences containing the 3′ UTR. When these sequences are introduced to the U937D cells, the synthesis of CK-B occurs. Possibly, exogenously applied 3′ UTR sequences bind to the translational repressor proteins, making them unavailable to bind to the CK-B mRNA. If these factors are not present on the CK-B mRNA, then synthesis of the CK-B protein can take place.

Section 12.4

42.   [Data Analysis Problem] In recent years, techniques have been developed to clone mammals through a process called nuclear transfer, in which the nucleus of a somatic cell is transferred to an egg cell from which the nuclear material has been removed. Research has demonstrated that when a nucleus from a *differentiated somatic cell* is transferred to an egg cell, only a small percentage of the resulting embryos complete development and many of those that do die shortly after birth. In contrast, when a nucleus from an *undifferentiated embryonic stem cell* is transferred into an egg cell, the percentage of embryos that complete development is significantly higher (W. M. Rideout, K. Eggan, and R. Jaenisch. 2001. *Science* 293:1095–1098). Why might the successful development of cloned embryos be higher when the nucleus transferred comes from an undifferentiated embryonic stem cell?

Solution:
During the process of development of a somatic cell, many genes that are not necessary for a particular cell type become silenced. Epigenetic changes to chromatin are often responsible for this type of gene silencing, and thus numerous epigenetic changes have occurred in the nuclear material from a differentiated somatic cell. These changes may silence genes that are necessary for successful development of all cell types in a developing embryo. The most likely explanation for failure of embryos created by nuclear transfer is the inability to "reprogram" epigenetic changes that occurred as the somatic cell developed. In contrast, nuclear material from undifferentiated embryonic stem cells has undergone less epigenetic reprogramming, allowing all genes to be activated when they are needed in development.

Chapter Thirteen: Gene Mutations, Transposable Elements, and DNA Repair

COMPREHENSION QUESTIONS

Section 13.1

1. What is the difference between a transition and a transversion? Which type of base substitution is usually more common?

    Solution:
    Transition mutations result from purine-to-purine or pyrimidine-to-pyrimidine base substitutions. Transversions result from purine-to-pyrimidine or pyrimidine-to- purine base substitutions. Transition mutations are more common because spontaneous mutations typically result in transition mutations rather than transversions.

2. Briefly describe expanding nucleotide repeats.

    Solution:
    Expanding nucleotide repeats result when a DNA insertion mutation increases the number of copies of a trinucleotide repeat sequence. The increase may be due to errors in replication or to unequal recombination.

3. What is the difference between a missense mutation and a nonsense mutation? A silent mutation and a neutral mutation?

    Solution:
    A base substitution that changes the sequence and the meaning of an mRNA codon, resulting in a different amino acid being inserted into a protein, is called a missense mutation. Nonsense mutations occur when a mutation replaces a sense codon with a stop (or nonsense) codon.

    A nucleotide substitution that changes the sequence of an mRNA codon, but not the meaning, is called a silent mutation. In neutral mutations, the sequence and the meaning of an mRNA codon are changed. However, the amino acid substitution has little or no effect on protein function.

4. Briefly describe two different ways in which intragenic suppressors can reverse the effects of mutations.

    Solution:
    Intragenic suppression is the result of second mutations within a gene that restore a wild-type phenotype. The suppressor mutations are located at different sites within the gene from the original mutation. One type of suppressor mutation restores the original phenotype by reverting the meaning of a previously mutated codon to that of the original

codon. The suppressor mutation occurs at a different position than the first mutation, which is still present within the codon. Intragenic suppression may also occur at two different locations within the same protein. If two regions of a protein interact, a mutation in one of these regions could disrupt that interaction. The suppressor mutation in the other region would restore the interaction. Finally, a frameshift mutation due to an insertion or deletion could be suppressed by a second insertion or deletion that restores the proper reading frame.

Section 13.2

5.   How do insertions and deletions arise?

Solution:
Strand slippage that occurs during DNA replication and unequal crossover events due to misalignment at repetitive sequences have been shown to cause deletions and additions of nucleotides to DNA molecules. Strand slippage results from the formation of small loops on either the template or the newly synthesized strand. If the loop forms on the template strand, then a deletion occurs. Loops formed on the newly synthesized strand result in insertions. If, during crossing over, a misalignment of the two strands at repetitive sequence occurs, then the resolution of the crossover will result in one DNA molecule containing an insertion and the other molecule containing a deletion.

6.   How do base analogs lead to mutations?

Solution:
Base analogs have structures similar to the nucleotides and can be incorporated into the DNA in the course of replication. Many analogs tend to mispair, which can lead to mutations. DNA replication is required for the base-analog-induced mutations to be incorporated into the DNA.

7.   What is the purpose of the Ames test? How are *his⁻* bacteria used in this test?

Solution:
The Ames test allows for rapid and inexpensive detection of potentially carcinogenic compounds using bacteria. The majority of carcinogenic compounds result in damage to DNA and are mutagens. The reversion of *his⁻* bacteria to *his⁺* is used to detect the mutagenic potential of the compound being tested.

Section 13.3

8.   What general characteristics are found in many transposable elements?

Solution:
Most transposable elements have terminal inverted repeats and are flanked by short direct repeats that are generated at insertion sites during the transposition process. Many also

contain a gene encoded with one of the enzymes necessary for transposition (transposase or reverse transcriptase).

9.   Describe the differences between replicative and nonreplicative transposition.

Solution:
Replicative transposons use a copy-and-paste mechanism in which the transposon is replicated and inserted in a new location, leaving the original transposon in place. Nonreplicative transposons use a cut-and-paste mechanism in which the original transposon is excised and moved to a new location.

10.   What is a retrotransposon and how does it move?

Solution:
A retrotransposon is a transposable element that relocates through an RNA intermediate. First, it is transcribed into RNA. Then, a reverse transcriptase encoded by the retrotransposon reverse transcribes the RNA template into a DNA copy of the transposon, which then integrates into a new location in the host genome.

Section 13.4

11.   List at least three different types of DNA repair and briefly explain how each is carried out.

Solution:
(1) Mismatch repair: Replication errors that result from base-pair mismatches. Mismatch-repair enzymes recognize distortions in the DNA structure due to mispairing and detect the newly synthesized strand by its lack of methylation. The distorted segment is excised, and DNA polymerase and DNA ligase fill in the gap.
(2) Direct repair: Repairs DNA damage by directly changing the damaged nucleotide back into its original structure.
(3) Base-excision repair: Excises the damaged base and then replaces the entire nucleotide.
(4) Nucleotide-excision repair: Relies on repair enzymes to recognize distortions of the DNA double-helix. These enzymes excise a damaged region by cleaving phosphodiester bonds on either side of the damaged region. The gap created by the excision is filled by DNA polymerase.

APPLICATION QUESTIONS AND PROBLEMS

Section 13.1

12.   A codon that specifies the amino acid Gly undergoes a single-base substitution to become a nonsense mutation. In accord with the genetic code given in **Figure 11.5**, is this mutation a transition or a transversion? At which position of the codon does the mutation occur?

Solution:
Transversion at the first position:

GGA → UGA

*13. Refer to the genetic code in **Figure 11.5** to answer the following questions:

   a. If a single transition occurs in a codon that specifies Phe, what amino acids can be specified by the mutated sequence?

   Solution:
   Two codons can encode for Phe, UUU, and UUC. A single transition could occur at each of the positions of the codon resulting in different meanings.

Original codon	Mutated codon (amino acid encoded)
UUU	CUU (Leu), UCU (Ser), UUC (Phe)
UUC	CUC (Ser), UCU (Ser), UUU (Phe)

   b. If a single transversion occurs in a codon that specifies Phe, what amino acids can be specified by the mutated sequence?

   Solution:

Original codon	Mutated codon (amino acid encoded)
UUU	AUU (Ile), UAU (Tyr), UUA (Leu), GUU (Val), UGU (Cys), UUG (Leu)
UUC	AUC (Ile), UAC (Tyr), UUA (Leu), GUC (Val), UGC (Cys), UUG (Leu)

   c. If a single transition occurs in a codon that specifies Leu, what amino acids can be specified by the mutated sequence?

   Solution:

Original codon	Mutated codon (amino acid encoded)
CUU	UUU (Phe), CCU (Pro), CUC (Leu)
CUC	UUC (Phe), CCC (Pro), CUG (Leu)
CUA	UUA (Leu), CCA (Pro), CUG (Leu)
CUG	UUG (Leu), CCG (Pro), CUA (Leu)
UUG	CUG (Leu), UCG (Ser), UUA (Ser)
UUA	CUA (Leu), UCG (Ser), UUG (Leu)

d. If a single transversion occurs in a codon that specifies Leu, what amino acids can be specified by the mutated sequence?

Solution:

Original codon	Mutated codon (amino acid encoded)
UUA	AUA (Met), UAA (Stop), UUU (Phe), GUA (Val), UGA (Stop), UUC (Phe)
UUG	AUG (Met), UAG (Stop), UUU (Phe), GUG (Val), UGG (Trp), UUC (Phe)
CUU	GUU (Val), CGU (Arg), CUG (Leu), AUU (Ile),
CUC	AUC (Ile), CAC (His), CUA (Leu), GUC (Val), CGC (Arg), CUG (Leu)
CUA	AUA (Ile), CAA (Gln), CUC (Leu), GUA (Val), CGA (Arg), CUG (Leu)
CUG	AUG (Met), CAG (Gln), CUC (Leu), GUG (Val), CGG (Arg), CUU (Leu)

14. Hemoglobin is a complex protein that contains four polypeptide chains. The normal hemoglobin found in adults—called adult hemoglobin—consists of two alpha and two beta polypeptide chains, which are encoded by different loci. Sickle-cell hemoglobin, which causes sickle-cell anemia, arises from a mutation in the beta chain of adult hemoglobin. Adult hemoglobin and sickle-cell hemoglobin differ in a single amino acid: the sixth amino acid from one end in adult hemoglobin is glutamic acid, whereas sickle-cell hemoglobin has valine at this position. After consulting the genetic code provided in **Figure 11.5**, indicate the type and location of the mutation that gave rise to sickle-cell anemia.

Solution:
There are two possible codons for glutamic acid, GAA and GAG. Single-base substitutions at the second position in both codons can produce codons that encode valine:

    GAA→ GUA (Val)
    GAG→ GUG (Val)

Both substitutions are transversions. However, in the gene encoding the beta chain of hemoglobin, the GAG codon is the wild-type codon and the mutated GUG codon results in the sickle-cell phenotype.

*15. The following nucleotide sequence is found on the template strand of DNA. First, determine the amino acids of the protein encoded by this sequence by using the genetic code provided in **Figure 11.5**. Then, give the altered amino acid sequence of the protein that will be found in each of the following mutations:

Sequence of DNA template: 3′–TAC TGG CCG TTA GTT GAT ATA ACT–5′
　　　　　Nucleotide number → 1   24

Solution:
mRNA sequence: 5′–AUG ACC GGC AAU CAA CUA UAU UGA–3′
amino acid sequence: Amino–Met Thr Gly Asn Gln Leu Tyr Stop–Carboxyl

a. Mutant 1: A transition at nucleotide 11
Amino–Met-Thr-Gly-**Ser**-Gln-Leu-Tyr-Stop–carboxyl
b. Mutant 2: A transition at nucleotide 13
Amino–Met Thr Gly Asn-**STOP**–carboxyl
c. Mutant 3: A one-nucleotide deletion at nucleotide 7
Amino–Met-Thr-**Ala-Ile-Asn-Tyr-Ile**–carboxyl
d. Mutant 4: A T→A transversion at nucleotide 15
Amino–Met-Thr-Gly-Asn-**His**-Leu-Tyr-Stop–carboxyl
e. Mutant 5: An addition of TGG after nucleotide 6
Amino–Met-Thr-**Thr**-Gly-Asn-Gln-Leu-Tyr-Stop–carboxyl
f. Mutant 6: A transition at nucleotide 9
Amino–Met-Thr-**Gly**-Asn-Gln-Leu-Tyr-Stop–carboxyl

16. Draw a hairpin turn like that shown in **Figure 13.5** for the repeated sequence found in fragile-X syndrome (see **Table 13.1**).

Solution:

```
      1                      6   7
      GCCGCCGCCGCCGCCGCCGCCGCCGCC
      CGGCGGCGG  CG    CGGCGGCGGCGG
      1          G   G  11
                  GC
                  CG
                 G  G
                  GC
                  CG
                 G  G
                  GC
                C    G
                  G

                  7
```

*17. A polypeptide has the following amino acid sequence:
         Met-Ser-Pro-Arg-Leu-Glu-Gly
The amino acid sequence of this polypeptide was determined in a series of mutants listed in parts *a* through *e*. For each mutant, indicate the type of change that occurred in the DNA (single-base substitution, insertion, deletion) and the phenotypic effect of the mutation (nonsense mutation, missense mutation, frameshift, etc.).

a. Mutant 1: Met-Ser-Ser-Arg-Leu-Glu-Gly

Solution:
A missense mutation has occurred resulting in the substitution of Ser for Pro in the protein. The change is most likely due to a single-base substitution in the Ser codon resulting in the production of a Pro codon. Four of the Ser codons can be changed to Pro codons by a single transition mutation.

Pro	**Ser**
CCU	UCU
CCC	UCC
CCA	UCA
CCG	UCG

b. Mutant 2: Met-Ser-Pro

Solution:
A single-base substitution has occurred in the Arg codon resulting in the formation of a stop codon. Two of the potential codons for Arg can be changed by single substitutions to stop codons. The phenotypic effect is a nonsense mutation.

Arg	**Stop**	
CGA	UGA	transition mutation
AGA	UGA	transversion mutation

c. Mutant 3: Met-Ser-Pro-Asp-Trp-Arg-Asp-Lys

Solution:
The deletion of a single nucleotide at the first position in the Arg codon (most likely CGA) has resulted in a frameshift mutation in which the mRNA is read in a different frame, producing a different amino acid sequence for the protein.

d. Mutant 4: Met-Ser-Pro-Glu-Gly

Solution:
A six-base-pair deletion has occurred, resulting in the elimination of two amino acids (Arg and Leu) from the protein. The result is a truncated polypeptide chain.

e. Mutant 5: Met-Ser-Pro-Arg-Leu-Leu-Glu-Gly

Solution:
The addition or insertion of three nucleotides into the DNA sequence has resulted in the addition of a Leu codon to the polypeptide chain.

18.   A gene encodes a protein with the following amino acid sequence:
            Met-Trp-His-Arg-Ala-Ser-Phe.
       A mutation occurs in the gene. The mutant protein has the following amino acid sequence:
            Met-Trp-His-Ser-Ala-Ser-Phe.
       An intragenic suppressor restores the amino acid sequence to that of the original protein:
            Met-Trp-His-Arg-Ala-Ser-Phe.
       Give at least one example of base changes that could produce the original mutation and the
       intragenic suppressor. (Consult the genetic code in **Figure 11.5**.)

       Solution:
       Four of the six Arg codons could be mutated by a single-base substitution to produce a Ser
       codon. However, only two of the Arg codons mutated to form Ser codons could be
       subsequently mutated at a second position by a single-base substitution to regenerate the
       Arg codon. In both events, the mutations are transversions.

Original Arg codon	Ser codon	Restored Arg codon
CGU	AGU	AGG or AGA
CGC	AGC	AGG or AGA

Section 13.2

19.   The following nucleotide sequence is found in a short stretch of DNA:
            5'–ATGT–3'
            3'–TACA–5'

       If this sequence is treated with hydroxylamine, what sequences will result after
       replication?

       Solution:

Original sequence	Mutated sequence
5' ATGT 3'	5'–ATAT–3'
3'–TACA–5'	3'–TATA–5'

*20.   The following nucleotide sequence is found in a short stretch of DNA:
            5'–AG–3'
            3'–TC–5'

       a.   Give all the mutant sequences that can result from spontaneous depurination in this
             stretch of DNA.

             Solution:
             The strand contains two purines, adenine, and guanine. Because repair of depurination
             typically results in adenine being substituted for the missing purine, only the loss of
             the guanine by depurination will result in a mutant sequence.

$$5'-\text{A}\mathbf{G}-3' \quad \text{to} \quad 5'-\text{A}\mathbf{A}-3'$$
$$3'-\text{T}\mathbf{C}-5' \qquad\qquad 3'-\text{T}\mathbf{T}-5'$$

b. Give all the mutant sequences that can result from spontaneous deamination in this stretch of DNA.

Solution:
Deamination of guanine, cytosine, and adenine can occur. However, the deamination of only cytosine and adenine are likely to result in mutant sequences because the deamination products can form improper base pairs. The deamination of guanine does not pair with thymine but can still form two hydrogen bonds with cytosine; thus no change will occur.

$$5'-\text{A}\text{G}-3' \qquad\qquad \text{if A is deaminated, then} \qquad 5'-\text{G}\text{G}-3'$$
$$3'-\text{T}\text{C}-5' \qquad\qquad\qquad\qquad\qquad\qquad\qquad 3'-\text{C}\text{C}-5'$$

$$5'-\text{A}\text{G}-3' \qquad\qquad \text{if C is deaminated, then} \qquad 5'-\text{A}\text{A}-3'$$
$$3'-\text{T}\text{C}-5' \qquad\qquad\qquad\qquad\qquad\qquad\qquad 3'-\text{T}\text{T}-5'$$

21. [Data Analysis Problem] Mary Alexander studied the effects of radiation on mutation rates in the sperm of *Drosophila melanogaster*. She irradiated *Drosophila* larvae with either 3000 roentgens (r) or 3975 r, collected the adult males that developed from irradiated larvae, mated them with unirradiated females that were homozygous for recessive alleles at 8 loci. She then counted the number of F_1 flies that carried a new mutation at each locus. All mutant flies that appeared were used in subsequent crosses to determine if their mutant phenotypes were genetic. For the roughoid locus, she obtained the following results (M. L. Alexander. 1954. *Genetics* 39:409–428):

Group	Number of offspring	Offspring with a mutation at the *roughoid* locus
Control (0 r)	45,504	0
Irradiated (3000 r)	49,512	5
Irradiated (3975 r)	50,159	16

a. Calculate the mutation rates at the *roughoid* locus of the control group and the two groups of irradiated flies.

Solution:
Because the females are homozygous for recessive alleles, any new mutation in a sperm cell will result in an offspring with a mutant phenotype; thus the mutation rate per gamete is simply the number of mutant offspring divided by the total number of offspring examined. Because no mutations were detected in the control group, the mutation rate per gamete must be less than 1/45,504 or 2.2×10^{-5}. The appearance of 5 mutations among 49,512 offspring gives a mutation rate of $5/49,512 = 1.0 \times 10^{-4}$ for the flies irradiated at 3000 r. The appearance of 16 mutations among 50,159 offspring gives a mutation rate of $16/50,159 = 3.19 \times 10^{-4}$ for the flies irradiated at 3975 r.

b. On the basis of these data, do you think radiation has any effect on mutation? Explain your answer.

Solution:
Yes. The radiation increased the mutation rate and the flies receiving more radiation (higher roentgens) produced a higher rate of mutant offspring.

22. What conclusion would you draw if the number of bacterial colonies in **Figure 13.21** were the same on the control plate and the treatment plate? Explain your reasoning.

Solution:
The chemical tested is not mutagenic and likely not carcinogenic. The number of colonies on each plate represents the number of bacterial cells that underwent a mutation. Because the number of cells undergoing mutation on the control plate, without the tested chemical, was the same as the number undergoing mutation on the plate treated with the chemical, there is no evidence that the chemical elevates the mutation rate or is potentially carcinogenic.

Section 13.3

*23. A particular transposable element generates flanking direct repeats that are 4 bp long. Give the sequence that will be found on both sides of the transposable element if this transposable element inserts at the position indicated on each of the following sequences:

Solution:

a. 5′—ATTCGAAC**TGAC**(transposable element)**TGAC**CGATCA—3′

b. 5′—ATT**CGAA**(transposable element)**CGAA**CTGACCGATCA—3′

24. What factor determines the length of the flanking direct repeats that are produced in a transposition?

Solution:
The number of base pairs between the staggered single-strand nicks made at the target site by the transposase.

25. Zidovudine (AZT) is a drug used to treat patients with AIDS. AZT works by blocking the reverse transcriptase enzyme used by human immunodeficiency virus (HIV), the causative agent of AIDS. Do you expect that AZT would have any effect on transposable elements? If so, what type of transposable elements would be affected and what would be the most likely effect?

Solution:
AZT should affect retrotransposons because they transpose through an RNA intermediate that is reverse transcribed to DNA by reverse transcriptase. If endogenous reverse transcriptases in human cells have similar sensitivity to AZT as HIV reverse transcriptase, then AZT should inhibit retrotransposons.

26.   A transposable element is found to encode a reverse-transcriptase enzyme. On the basis of this information, what conclusions can you make about the most likely method of transposition of this element?

Solution:
Like other retrotransposons, this element probably has long terminal direct repeats and transposes through an RNA intermediate that is reverse transcribed to DNA.

Section 13.4

*27.   A plant breeder wants to isolate mutants in tomatoes that are defective in DNA repair. However, this breeder does not have the expertise or equipment to study enzymes in DNA-repair systems. How can the breeder identify tomato plants that are deficient in DNA repair? What are the traits to look for?

Solution:
By looking for plants that have increased levels of mutations either in their germ-line or somatic tissues. Potentially mutant plants may have been exposed to standard mutagens that damage DNA. If they are defective in DNA repair, they should have higher rates of mutation. For example, tomato plants with defective DNA-repair systems should have an increased mutation rate when exposed to high levels of ultraviolet light. Therefore, they need to be grown in an environment that has lower levels of sunlight.

CHALLENGE QUESTIONS

Section 13.1

28.   [Data Analysis Problem] Robert Bost and Richard Cribbs studied a strain of *E. coli* (*araB14*) that possessed a nonsense mutation in the structural gene that encodes L-ribulokinase, an enzyme that allows the bacteria to metabolize the sugar arabinose (R. Bost and R. Cribbs. 1969. *Genetics* 62:1–8). From the *araB14* strain, they isolated some bacteria that possessed mutations that caused the bacteria to revert back to wild type. Genetic analysis of these revertants showed that they possessed two different suppressor mutations. One suppressor mutation (*R1*) was linked to the original mutation in the L-ribulokinase and probably occurred at the same locus. By itself, this mutation allowed the production of L-ribulokinase, but the enzyme was not as effective in metabolizing arabinose as the enzyme encoded by the wild-type allele. The second suppressor mutation (*Su*^{B}) was not linked to the original mutation. In conjunction with the *R1* mutation, *Su*^{B} allowed the production of L-ribulokinase, but *Su*^{B} by itself was not able to suppress the original mutation.

a. On the basis of this information, are the *R1* and *Su*^B mutations intragenic suppressors or intergenic suppressors? Explain your reasoning.

Solution:
R1 is an intragenic suppressor. As the studies indicated, it likely occurred within the same locus as the *araB14* mutation. *Su*^B is an intergenic suppressor because it was not linked to the original mutation and occurred at a different locus.

b. Propose an explanation for how *R1* and *Su*^B restore the ability of *araB14* to metabolize arabinose and why *Su*^B is able to more fully restore the ability.

Solution:
Potentially, the *R1* mutation changed the nonsense codon found in *araB14* into a sense codon although not back to the original wild-type codon. The new sense codon in the *R1* mutation allows for the insertion of an amino acid and a full-length protein is now synthesized that contains a missense mutation. *Su*^B encodes a mutation in a *tRNA* anticodon that allows for another amino acid to be inserted at the *R1* mutant codon restoring more of the L-ribulokinase function.

29. Achondroplasia is an autosomal dominant disorder characterized by disproportionate short stature—the legs and arms are short compared with the head and trunk. The disorder is due to a base substitution in the gene, located on the short arm of chromosome 4, for fibroblast growth factor receptor 3 (FGFR3).

Although achondroplasia is clearly inherited as an autosomal dominant trait, more than 80% of the people who have achondroplasia are born to parents with normal stature. This high percentage indicates that most cases are caused by newly arising mutations; these cases (not inherited from an affected parent) are referred to as sporadic. Findings from molecular studies have demonstrated that sporadic cases of achondroplasia are almost always caused by mutations inherited from the father (paternal mutations). In addition, the occurrence of achondroplasia is higher among older fathers; indeed, approximately 50% of children with achondroplasia are born to fathers older than 35 years of age. There is no association with maternal age. The mutation rate for achondroplasia (about 4×10^{-5} mutations per gamete) is high compared with those for other genetic disorders. Explain why most spontaneous mutations for achondroplasia are paternal in origin and why the occurrence of achondroplasia is higher among older fathers.

Solution:
In men, sperm cells are produced throughout much of their life. The cells responsible for the sperm production are called spermatogonia. These spermatogonia divide by mitosis to produce more spermatogonia and produce spermatocytes, cells that eventually will divide by meiosis to produce sperm cells. These continued cell divisions by the spermatogonia, particularly to produce more spermatogonia, could lead to an increased chance of mutations within the DNA of the spermatogonia cell. Essentially, the more cell divisions the greater the risk for mutation. Also some DNA sequences may be more susceptible to mutations than others. These locations are called hot spots. Potentially, the base

substitution occurs at a hot spot in the FGFR3 gene as more cell divisions occur. In addition, as men age, their exposure to environmental factors over time may increase mutation rates including that of the FGFR3 gene.

However, recent data do not support that the increase in mutations is due only to an increase in mutations as a result of spermatogonia mitosis. A second possibility is that the mutation may confer a positive benefit to the spermatogonia or, ultimately, to the sperm cells produced. Potentially, the mutated sperm have a higher survival rate than normal sperm cells leading to an increased risk of fertilization by the FGFR3 mutation containing sperm cells.

30.   *Ochre* and *amber* are two types of nonsense mutations. Before the genetic code was worked out, Sydney Brenner, Anthony O. Stretton, and Samuel Kaplan applied different types of mutagens to bacteriophages in an attempt to determine the bases present in the codons responsible for *amber* and *ochre* mutations. They knew that *ochre* and *amber* mutants were suppressed by different types of mutations, demonstrating that each is a different termination codon. They obtained the following results:
(1) A single-base substitution could convert an *ochre* mutation into an *amber* mutation.
(2) Hydroxylamine induced both *ochre* and *amber* mutations in wild-type phages.
(3) 2-Aminopurine caused *ochre* to mutate to *amber*.
(4) Hydroxylamine did not cause *ochre* to mutate to *amber*.

These data do not allow the complete nucleotide sequence of the *amber* and *ochre* codons to be worked out, but they do provide some information about the bases found in the nonsense mutations.

a. What conclusions about the bases found in the codons of *amber* and *ochre* mutations can be made from these observations?

Solution:
In considering the data, it is important to remember the mutagenic actions of hydroxylamine and 2-aminopurine. Hydroxylamine produces only GC to AT transition mutations. However, 2-aminopurine can produce both types of transitions, GC to AT and AT to GC.

Because hydroxylamine can be used to produce *amber* and *ochre* mutations from wild-type phages, then *amber* and *ochre* codons must contain uracil and/or adenine. The production of *amber* mutations from *ochre* codons by 2-aminopurine but not by hydroxylamine suggests that *ochre* mutations do contain adenine and uracil, whereas *amber* mutations contain guanine and also contain either adenine and/or uracil.

b. Of the three nonsense codons (UAA, UAG, UGA), which represents the *ochre* mutation?

Solution:
Based on the mutagenesis data, only the UAA codon matches the results for the ochre mutation. It is the only stop codon that does not contain guanine.

Chapter Fourteen: Molecular Genetic Analysis and Biotechnology

COMPREHENSION QUESTIONS

Section 14.2

1. What role do restriction enzymes play in bacteria? How do bacteria protect their own DNA from the action of restriction enzymes?

   Solution:
   Restriction enzymes cut foreign DNA, such as viral DNA, into fragments. Bacteria protect their own DNA by modifying bases, usually by methylation, at the recognition sites.

2. Explain how gel electrophoresis is used to separate DNA fragments of different lengths.

   Solution:
   Gel electrophoresis uses an electric field to drive negatively charged DNA molecules through a gel that acts as a molecular sieve. Shorter DNA molecules are less hindered by the agarose or polyacrylamide matrix and migrate faster than do longer DNA molecules.

3. Give three important characteristics of cloning vectors.

   Solution:
   (1) an origin of DNA replication so they can be maintained in a cell
   (2) a gene, such as an antibiotic-resistance gene, to select for cells that carry the vector
   (3) a unique restriction site or series of sites into which a foreign DNA molecule may be inserted

Section 14.3

4. Briefly explain how the polymerase chain reaction is used to amplify a specific DNA sequence.

   Solution:
   First, the double-stranded template DNA is denatured by high temperature. Then, synthetic oligonucleotide primers corresponding to the ends of the DNA sequence to be amplified are annealed to the single-stranded DNA template strands. These primers are extended by a thermostable DNA polymerase so that the target DNA sequence is duplicated. These steps are repeated 30 times or more. Each cycle of denaturation, primer annealing, and extension doubles the number of copies of the target sequence between the primers.

5. Briefly explain how an antibiotic-resistance gene and the *lacZ* gene can be used as markers to determine which cells contain a particular plasmid.

Solution:
Foreign DNAs are inserted into one of the unique restriction sites in the *lacZ* gene and transformed into *E. coli* cells. Transformed cells are plated on a medium containing the appropriate antibiotic to select for cells that carry the plasmid, an inducer of the *lac* operon, and X-gal, a substrate for β-galactosidase that turns blue when cleaved. Colonies that carry the plasmid without foreign DNA inserts will have intact *lacZ* genes, make functional β-galactosidase, cleave X-gal, and turn blue. Colonies that carry plasmid with foreign DNA inserts will not make functional β-galactosidase and will remain white.

Section 14.4

6.   How does a genomic library differ from a cDNA library?

Solution:
A genomic library is created by cloning fragments of chromosomal DNA into a cloning vector. Chromosomal DNA is randomly fragmented by shearing or by partial digestion with a restriction enzyme. A cDNA library is made from mRNA sequences. Cellular mRNAs are isolated and then reverse transcriptase is used to copy the mRNA sequences to cDNA, which are cloned into plasmid or phage vectors.

7.   Briefly explain how a gene can be isolated through positional cloning.

Solution:
The approximate location of a gene on a chromosome is identified by recombination or deletion mapping, with markers and deletions with known positions on the chromosome. All genes within this region are characterized to determine which gene has mutations that co-segregate with mutant phenotypes.

Section 14.5

8.   What is the purpose of the dideoxynucleoside triphosphates in the dideoxy sequencing reaction?

Solution:
Dideoxynucleoside triphosphates (ddNTPs) act as a substrate for DNA polymerase but cause termination of DNA synthesis when they are incorporated. Mixed with regular dNTPs, fluorescently labeled ddNTPs generate a series of DNA fragments that have terminated at every nucleotide position along the template DNA molecule being sequenced. These fragments can be separated by gel electrophoresis. Because each of the four ddNTPs carries a different fluorescent label, a laser detector can distinguish which base terminates each fragment. Reading the fragments from shorter to longer, an automated DNA sequencer can determine the sequence of the template DNA molecule.

9.   What is DNA fingerprinting? What types of sequences are examined in DNA
      fingerprinting?

      Solution:
      DNA fingerprinting detects genetic differences among people by using probes for highly
      variable regions (usually, microsatellites or short tandem repeats) of chromosomes.
      Typically, PCR is used to detect the microsatellites. Because people differ in the number
      of repeated sequences that they possess, the technique is used in the analysis of crimes, in
      paternity cases, and in identifying remains.

Section 14.6

10.   How does a reverse-genetics approach differ from a forward-genetics approach?

      Solution:
      Forward genetics begins with mutant phenotype and proceeds toward cloning and
      characterization of the DNA encoding the gene. Reverse genetics begins with a DNA
      sequence and then generates mutants to characterize the functions of the gene.

11.   What are knockout mice and for what are they used?

      Solution:
      A knockout mouse has a target gene disrupted or deleted ("knocked-out"). The offspring
      phenotype provides information about the function of the gene.

12.   How is RNA interference used in the analysis of gene function?

      Solution:
      RNA interference is one potential reverse-genetics approach to analyze gene function, by
      specifically repressing expression of that gene. Double-stranded RNA may be injected
      directly into a cell or organism, or the cell or organism may be genetically modified to
      express a double-stranded RNA molecule corresponding to the target gene.

Section 14.7

13.   What is gene therapy?

      Solution:
      Gene therapy is the correction of a defective gene by either gene replacement or the
      addition of a wild-type copy of the gene. For this to work, enough of the cells of the
      critically affected tissues or organs must be transformed with the functional copy of the
      gene to restore normal physiology.

APPLICATION QUESTIONS AND PROBLEMS

Section 14.2

14.   CRISPR-Cas9 was first developed as a molecular tool in 2012; during the next few years, its use in molecular biology exploded, as scientists around the world began applying it to many different research problems, and hundreds of research papers describing its application were published. Explain why CRISPR-Cas is such a powerful tool in molecular genetics.

Solution:
CRISPR-Cas technology is powerful, versatile, and relatively easy to use compared to previous technologies for genetic modification of organisms. CRISPR-Cas is readily programmable against different target sequences just by changing the sequence of the sgRNA. The sgRNAs are more specific than restriction enzymes because they recognize longer nucleotide sequences and thus have fewer off-target effects. CRISPR-Cas can work in any cell or organism where the sgRNA and Cas can be introduced, either by transfection or micro-injection. It can be used either to render genes nonfunctional via nonhomologous end joining or to introduce specific alterations via homologous recombination. Because the system works at high efficiencies, it does not require introduction of foreign DNA such as neomycin resistance for selection. Thus, molecular geneticists now have a powerful tool for precise genome editing, that is, for creation of genetically modified organisms that are not transgenic. The technique is applicable to a wide variety of species, from bacteria to fungi, plants, animals, and humans.

*15.   Suppose that a geneticist discovers a new restriction enzyme in the bacterium *Aeromonas ranidae*. This restriction enzyme is the first to be isolated from this bacterial species. Using the standard convention for abbreviating restriction enzymes, give this new restriction enzyme a name (for help, see the footnote to **Table 14.1**).

Solution:
The first three letters are taken from the genus and species name, and the Roman numeral indicates the order in which the enzyme was isolated. Therefore, the enzyme should be named **AraI**.

16.   How often, on average, would you expect a restriction endonuclease to cut a DNA molecule if the recognition sequence for the enzyme had 5 bp? (Assume that the four types of bases are equally likely to be found in the DNA and that the bases in a recognition sequence are independent.) How often would the endonuclease cut the DNA if the recognition sequence had 8 bp?

Solution:
Because DNA has four different bases, the frequency of any sequence of n bases is equal to $1/(4^n)$. A 5-bp recognition sequence will appear with a frequency of $1/(4^5)$, or once every 1024 bp. An 8-bp recognition sequence will appear with a frequency of $1/(4^8)$, or 65,536 bp.

*17. A microbiologist discovers a new restriction endonuclease. When DNA is digested by this enzyme, fragments that average 1,048,500 bp in length are produced. What is the most likely number of base pairs in the recognition sequence of this enzyme?

Solution:
Here, $4^n = 1,048,500$, so $n = 10$. A 10-bp recognition sequence is most likely.

18.   Will restriction sites for an enzyme that has 4 bp in its restriction site be closer together, farther apart, or similarly spaced, on average, compared with those of an enzyme that has 6 bp in its recognition sequence? Explain your reasoning.

Solution:
The restriction sites for an enzyme with a 4-bp recognition sequence should be spaced closer together than the sites for an enzyme with a 6-bp recognition sequence. The 4-bp recognition sequence will occur with an average frequency of once every $4^4 = 256$ bp, whereas the 6-bp recognition sequence will occur with an average frequency of once every $4^6 = 4096$ bp.

*19. About 60% of the base pairs in a human DNA molecule are AT. If the human genome has 3.2 billion base pairs of DNA, about how many times will the following restriction sites be present?

   a. *Bam*HI (recognition sequence is 5′—GGATCC—3′)
   b. *Eco*RI (recognition sequence is 5′—GAATTC—3′)
   c. *Hae*III (recognition sequence is 5′—GGCC—3′)

Solution:
We must first calculate the frequency of each base. Given that AT base pairs consist 60% of the DNA, we deduce that the frequency of A is 0.3 and frequency of T is 0.3. The GC base pairs must consist of 40% of the DNA; therefore, the frequency of G is 0.2 and the frequency of C is 0.2.

   a. *Bam*HI GGATCC is then $(0.2)(0.2)(0.3)(0.3)(0.2)(0.2) = 0.000144$
       $3,200,000,000(0.000144) = 460,800$ times

   b. *Eco*RI GAATTC $= (0.2)(0.3)(0.3)(0.3)(0.3)(0.2) = 0.000324$
       $3,200,000,000(0.000324) = 1,036,800$ times

   c. *Hae*III GGCC $= (0.2)(0.2)(0.2)(0.2) = 0.0016$
       $3,200,000,000(0.0016) = 5,120,000$ times

*20. A linear piece of DNA has the following *Eco*RI restriction sites.

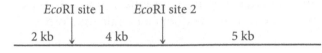

a. This piece of DNA is cut by *Eco*RI, the resulting fragments are separated by gel electrophoresis, and the gel is stained with ethidium bromide. Draw a picture of the bands that will appear on the gel.

b. If a mutation that alters *Eco*RI site 1 occurs in this piece of DNA, how will the banding pattern on the gel differ from the one that you drew in part *a*?

c. If mutations that alter *Eco*RI sites 1 and 2 occur in this piece of DNA, how will the banding pattern on the gel differ from the one that you drew in part *a*?

d. If 1000 bp of DNA were inserted between the two restriction sites, how would the banding pattern on the gel differ from the one you drew in part *a*?

e. If 500 bp of DNA between the two restriction sites were deleted, how would the banding pattern on the gel differ from the one that you drew in part *a*?

Solution:

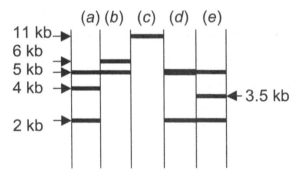

Section 14.3

*21. Which vectors (plasmid, phage λ, cosmid, bacterial artificial chromosome) can be used to clone a continuous fragment of DNA with the following lengths?

a. 4 kb
b. 20 kb
c. 35 kb
d. 100 kb

Solution:
a. 4 kb—plasmid
b. 20 kb—phage λ
c. 35 kb—cosmid
d. 100 kb—bacterial artificial chromosome

22.   A geneticist uses a plasmid for cloning that has the *lacZ* gene and a gene that confers resistance to penicillin. The geneticist inserts a piece of foreign DNA into a restriction site that is located within the *lacZ* gene and uses the plasmid to transform bacteria. Explain how the geneticist can identify bacteria that contain a copy of a plasmid with the foreign DNA.

Solution:
The geneticist should plate the bacteria on agar medium containing penicillin to select for cells that have taken up the plasmid. The medium should also have X-gal and an inducer of the lac operon, such as IPTG or even lactose. Cells that have taken up a plasmid without foreign DNA will have an intact *lacZ* gene, produce functional β-galactosidase, and cleave X-gal to make a blue dye. These colonies will turn blue. In contrast, cells that have taken up a plasmid containing a foreign DNA inserted into the *lacZ* gene will be unable to make functional β-galactosidase. These colonies will be white.

Section 14.4

23.   Suppose that you have just graduated from college and have started working at a biotechnology firm. Your first assignment is to clone the pig gene for the hormone prolactin. Assume that the pig gene for prolactin has not yet been isolated, sequenced, or mapped; however, the mouse gene for prolactin has been cloned and the amino acid sequence of mouse prolactin is known. Briefly explain two different strategies that you might use to find and clone the pig gene for prolactin.

Solution:
One strategy would be to use the mouse gene for prolactin as a probe to find the homologous pig gene from a pig genomic or cDNA library.

A second strategy would be to use the amino acid sequence of mouse prolactin to design degenerate oligonucleotides as hybridization probes to screen a pig DNA library.

Yet a third strategy would be to use the amino acid sequence of mouse prolactin to design a pair of degenerate oligonucleotide PCR primers for PCR amplification of the pig gene for prolactin.

Section 14.5

25.   Suppose that you want to sequence the following DNA fragment:

5′—TCCCGGGAAA-primer site—3′

You first use PCR to amplify the fragment, so that there is sufficient DNA for sequencing. You carry out dideoxy sequencing of the fragment. You then separate the products of the polymerization reactions by gel electrophoresis. Draw the bands that should appear on the gel from the four sequencing reactions.

Solution:

Note that, if the primer is labeled, bands will appear on all four lanes at the 5′ terminus of the DNA template fragment, as shown in the upper figure; the chain will terminate in all four reactions at this position because this is the end of the template. Thus, the 5′-end nucleotide cannot be determined by looking at bands in the sequencing gel. If the dideoxynucleotides are labeled, then the labeled band will appear in only the ddA lane in the uppermost position, as shown in the lower figure.

*26.   Suppose that you are given a short fragment of DNA to sequence. You amplify the fragment with PCR and set up a series of four dideoxy reactions. You then separate the products of the reactions by gel electrophoresis and obtain the following banding pattern:

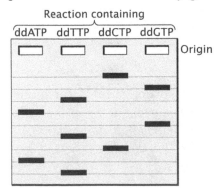

Write out the base sequence of the original fragment that you were given.
Original sequence: 5'—_____—3'

Solution:
Sequence of the newly synthesized strand from reading the gel: 5'–TACTGATGCN–3'
Original template strand sequence, complementary to newly synthesized strand:
5'–NGCATCAGTA–3'. The base at the 5' end (N) cannot be determined because the chain
stops in all four lanes.

27.   [Data Analysis Problem] The picture below is a sequencing gel from the original study
       that first sequenced the cystic fibrosis gene (J. R. Riordan et al. 1989. *Science*
       245:1066–1073). From the picture, determine the sequence of the normal copy of the gene
       and the sequence of the mutated copy of the gene. Identify the location of the mutation that
       causes cystic fibrosis (CF). Hint: The CF mutation is a 3-bp deletion.

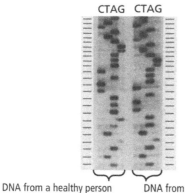

DNA from a healthy person   DNA from a person with CF

Solution:
Normal: 5'—ATAGTAGAAACCACAAAGGATACTA—3'
CF: 5'—ATAGTA[   ]ACCACAAAGGATACTACTT—3'
The empty bracket in the CF sequence denotes the location of 3 bases deleted in the CF allele.

Section 14.6

28.   You have discovered a gene in mice that is similar to a gene in yeast. How might you
       determine whether this gene is essential for development in mice?

       Solution:
       This gene must first be cloned, possibly by using the yeast gene as a probe to screen a
       mouse genomic DNA library. The cloned gene is then engineered to replace a substantial
       portion of the protein-coding sequence with the neo gene. This construct is then introduced
       into mouse embryonic stem cells, which are transferred to the uterus of a pseudopregnant
       mouse. The progeny are tested for the presence of the knockout allele, and those having
       the knockout allele are interbred. If the gene is essential for embryonic development, no
       homozygous knockout mice will be born. The arrested or spontaneously aborted fetuses
       can then be examined to determine how development has gone awry in fetuses that are
       homozygous for the knockout allele.

CHALLENGE QUESTIONS

Section 14.6

29.   Suppose that you are hired by a biotechnology firm to produce a strain of giant fruit flies by using recombinant DNA technology so that genetics students will not be forced to strain their eyes when looking at tiny flies. You go to the library and learn that growth in fruit flies is normally inhibited by a hormone called shorty substance P (SSP). You decide that you can produce giant fruit flies if you can somehow turn off the production of SSP. Shorty substance P is synthesized from a compound called XSP in a single-step reaction catalyzed by the enzyme runtase:

$$XSP \xrightarrow{\text{Runtase}} SSP$$

A researcher has already isolated cDNA for runtase and has sequenced it, but the location of the runtase gene in the *Drosophila* genome is unknown. In attempting to devise a strategy for turning off the production of SSP and producing giant flies by using standard recombinant DNA techniques, you discover that deleting, inactivating, or otherwise mutating this DNA sequence in *Drosophila* turns out to be extremely difficult. Therefore, you must restrict your genetic engineering to gene augmentation (adding new genes to cells). Describe the methods that you will use to turn off SSP and produce giant flies by using recombinant DNA technology.

Solution:
One possible solution is to create a gene for expression of dsRNA for runtase in the flies. Given the DNA sequence of runtase, one could create a vector with inverted repeats of the cDNA for runtase downstream of a strong promoter, so that hairpin dsRNA molecules of the runtase gene are produced in abundance. These dsRNA molecules will invoke the RNAi pathway to suppress runtase expression. Another route is to design a ribozyme whose targeting sequences are complementary to runtase mRNA. A DNA construct that would express a ribozyme can be transformed into the cell. The expressed ribozyme would then selectively target and cleave runtase mRNA.

Chapter Fifteen: Genomics and Proteomics

COMPREHENSION QUESTIONS

Section 15.1

1. What is the difference between a genetic map and a physical map? Which generally has higher resolution and accuracy and why?

   Solution:
   A genetic map locates genes or markers on the basis of genetic recombination frequencies. A physical map locates genes or markers on the basis of physical lengths of DNA sequence. Because recombination frequencies vary from one region of the chromosome to another, genetic maps are approximate. Genetic maps also have lower resolution because recombination is difficult to observe between loci that are very close to each other. Physical maps based on DNA sequences or restriction maps have much greater accuracy and resolution, down to a single base pair of DNA sequence.

2. What is the difference between a map-based approach to sequencing a whole genome and a whole-genome shotgun approach?

   Solution:
   The map-based approach first assembles large clones into contigs on the basis of genetic and physical maps and then selects clones for sequencing. The whole genome shotgun approach breaks the genome into short sequences—typically, from 600 to 700 bp—and then, with the use of powerful computers, assembles them into contigs on the basis of sequence overlap.

3. What is a single-nucleotide polymorphism (SNP)? How are SNPs used in genomic studies?

   Solution:
   SNPs are single base-pair differences in the sequence of a particular region of DNA from one individual compared to another of the same species or population. SNPs are useful as molecular markers for mapping and pedigree analysis and may themselves be associated with phenotypic differences.

4. What is a haplotype?

   Solution:
   A haplotype is a particular set of neighboring SNPs or other DNA polymorphisms observed on a single chromosome or chromosome region. They tend to be inherited together as a set because of linkage. Meiotic recombination within the chromosomal region can split the haplotype and create new recombinant haplotypes.

5.   How is a genome-wide association study carried out?

Solution:
Genome-wide association studies survey numerous genetic polymorphisms scattered throughout the genome, particularly single-nucleotide polymorphisms (SNPs), for linkage to diseases or to other traits. Hundreds or thousands of individuals are each typed for hundreds of thousands of SNPs. SNPs that show strong statistical linkage to a particular trait indicate that a gene that contributes to the trait is located near the SNPs. Because entire genomes are surveyed in these individuals, genome-wide association studies can reveal multiple genes that contribute incrementally to complex, quantitative traits.

6.   Give some examples of important findings from metagenomic studies.

Solution:
Metagenomic studies, also known as environmental genomics, have been especially useful for studying microbial communities, as the vast majority of microbial species cannot be cultured in the laboratory, precluding their isolation, identification, and characterization as individual species or strains. Analysis of ocean samples led to the discovery and characterization of proteorhodopsin proteins, light-driven proton pumps that represent a major driver of energy flux in the world's oceans. Analysis of gut bacteria revealed that obese and lean people have significant differences in their dominant bacterial species, as do obese and lean mice. These findings suggest that gut microflora may play a role in obesity. In general, metagenomics yields important insights into microbial communities in varied environments.

Section 15.2

7.   What is a microarray? How can it be used to obtain information about gene function?

Solution:
A microarray consists of thousands of DNA fragments spotted onto glass slides in an ordered grid (gene chips) or even proteins or peptides arrayed onto glass slides (protein chips). The identity of the DNA or peptide at each location is known. Gene chips are typically used in hybridization experiments with labeled mRNAs or cDNAs to survey the levels of transcript accumulation for thousands of genes, or even whole genomes, at one time. Peptide or protein chips can be used to identify protein–protein interactions or enzymatic activities or other properties of proteins.

Section 15.3

8.   What is the relation between genome size and gene number in prokaryotes?

Solution:
In prokaryotes, the gene number is proportional to the genome size because most of the genome encodes proteins.

9.   DNA content varies considerably among different multicellular organisms. Is this variation closely related to the number of genes and the complexity of the organism? If not, what accounts for the variation?

Solution:
This question is almost a philosophical one because "complexity" of an organism is not well-defined and thus difficult to quantify. However, we do know that the genomic DNA content can vary widely among related species, so there appears to be little relation between the "complexity" of an organism, the number of genes, and the DNA content. Large differences in DNA content may arise from differences in the frequency and size of introns, the abundance of DNA derived from transposable elements, and duplication of the whole or substantial parts of the genome in the evolutionary history of the species.

10.   What was the focus of the ENCODE project?

Solution:
The focus of the ENCODE project was to determine whether noncoding DNA had any function.

Section 15.4

11.   How does proteomics differ from genomics?

Solution:
Genomics is the analysis of whole genome DNA sequences and their organization, expression, and function, whereas proteomics focuses on the complete set of proteins made by an organism. Unlike the genome, which is constant from cell to cell, the proteome varies within a species from cell type to cell type, with stage of development, and with time in response to signals and other environmental stimuli. Proteins also undergo numerous modifications that affect protein localization and function.

12.   How is mass spectrometry used to identify proteins in a cell?

Solution:
Protein fragments generated by protease digestion are separated by mass and charge. Computer algorithms compare the mass profiles of these fragments with databases of known fragments of proteins or predicted fragments in a genome sequence.

APPLICATION QUESTIONS AND PROBLEMS

Section 15.1

13.   A 22-kb piece of DNA has the following restriction sites:

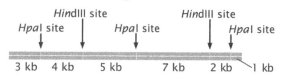

A batch of this DNA is first fully digested by *Hpa*I alone, then another batch is fully digested by *Hind*III alone, and, finally, a third batch is fully digested by *Hpa*I and *Hind*III together. The fragments resulting from each of the three digestions are placed in separate wells of an agarose gel, separated by gel electrophoresis, and stained by ethidium bromide. Draw the bands as they would appear on the gel.

Solution:

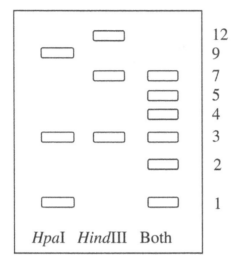

*14.   A linear piece of DNA that is 14 kb long is cut first by *Eco*RI alone, then by *Sma*I alone, and, finally, by *Eco*RI and *Sma*I together. The following results are obtained:

Digestion by EcoRI alone	Digestion by SmaI alone	Digestion by EcoRI and SmaI
3-kb fragment	7-kb fragment	2-kb fragment
5-kb fragment	7-kb fragment	3-kb fragment
6-kb fragment		4-kb fragment
		5-kb fragment

Draw a map of the *Eco*RI and *Sma*I restriction sites on this 14-kb piece of DNA, indicating the relative positions of the restriction sites and the distances between them.

Solution:
We know that *Sma*I cuts only once, in the middle of this piece of DNA, at 7 kb. *Eco*RI cuts twice. Comparing the *Eco*RI digest to the double digest, we see that neither the 3-kb nor the 5-kb fragments is cut by *Sma*I; only the 6-kb *Eco*RI fragment is cut by *Sma*I to 2-kb and 4-kb fragments. Therefore, the 6-kb *Eco*RI fragment is in the middle, and the 3-kb and 5-kb *Eco*RI fragments are at the ends.

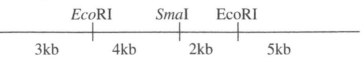

*15. A linear piece of DNA was broken into random, overlapping fragments and each fragment was sequenced. The sequence of each fragment is shown below.

Fragment 1: 5′—TAGTTAAAAC—3′

Fragment 2: 5′—ACCGCAATACCCTAGTTAAA—3′

Fragment 3: 5′—CCCTAGTTAAAAC—3′

Fragment 4: 5′—ACCGCAATACCCTAGTT—3′

Fragment 5: 5′—ACCGCAATACCCTAGTTAAA—3′

Fragment 6: 5′—ATTTACCGCAAT—3′

On the basis of overlap in sequence, assemble the fragments into a contig.

Solution:
We align the sequences with overlaps as follows:

```
              1: 5′—TAGTTAAAAC—3′
          2: 5′—ACCGCAATACCCTAGTTAAA—3′
             3: 5′—CCCTAGTTAAAAC—3′
          4: 5′—ACCGCAATACCCTAGTT—3′
          5: 5′—ACCGCAATACCCTAGTTAAA—3′
      6: 5′—ATTTACCGCAAT—3′
```

Sequence of original fragment: 5′—ATTTACCGCAATACCCTAGTTAAAAC—3′

16. In recent years, honeybee colonies throughout North America have been decimated by colony collapse disorder (CCD), which results in the rapid death of worker bees. First noticed by bee keepers in 2004, the disorder has been responsible for the loss of 50% to 90% of beekeeping operations in the United States. Evidence suggests that CCD is caused by a pathogen. Diana Cox-Foster and her colleagues (2007. *Science* 318:283–287) used a metagenomic approach to try to identify the causative agent of CCD. They isolated DNA from normal honeybee hives and from hives that had experienced CCD. A number of different bacteria, fungi, and viruses were identified in the metagenomic analysis. The following table gives the percentages of CCD hives and non-CCD hives that tested positive for four potential pathogens identified in the metagenomic analysis. On the basis of these data, which potential pathogen appears most likely to be responsible for CCD? Explain your reasoning. Do these data prove that this pathogen is the cause of CCD? Explain.

Virus	% of CCD hives infected ($n = 30$)	% of non-CCD hives infected ($n = 21$)
Israeli acute paralysis virus	83.3%	4.8%
Kashmir bee virus	100%	76.2%
Nosema apis	90%	47.6%
Nosema cernae	100%	80.8%

Solution:
The biggest difference in percent of infection of CCD and non-CCD colonies is for Israeli acute paralysis virus: 83% of CCD infected colonies had this virus versus only 4.8% of non-CCD colonies. However, this does not prove that Israeli acute paralysis virus is the cause of CCD. It is possible that CCD is caused by some other factor, but that once infected, CCD colonies are more vulnerable to infection by Israeli acute paralysis virus. Causation could be established by infecting non-CCD colonies with the virus and seeing if CCD develops.

17.   James Noonan and his colleagues (2005. *Science* 309:597–599) set out to study genome sequences of an extinct species of cave bear. They extracted DNA from 40,000-year-old bones from a cave bear. They then used a metagenomic approach to isolate, identify, and sequence the cave-bear DNA. Why did they use a metagenomic approach when their objective was to sequence the genome of one species (the cave bear)?

Solution:
The cave-bear bones were heavily contaminated with bacterial, fungal, and eukaryotic DNA. The researchers used metagenomics to sequence all of the DNA. They then compared the sequences obtained with modern bear DNA to identify which DNA fragments were from the cave bear.

Section 15.2

*18.   Microarrays can be used to determine the levels of gene expression. In one type of microarray, hybridization of the red (experimental) and green (control) cDNAs is proportional to the relative amounts of mRNA in the samples. Red indicates the overexpression of a gene and green indicates the underexpression of a gene in the experimental cells relative to the control cells, yellow indicates equal expression in experimental and control cells, and no color indicates no expression in either experimental or control cells.

In one experiment, mRNA from a strain of antibiotic-resistant bacteria (experimental cells) is converted into cDNA and labeled with red fluorescent nucleotides; mRNA from a nonresistant strain of the same bacteria (control cells) is converted into cDNA and labeled with green fluorescent nucleotides. The cDNAs from the resistant and nonresistant cells are mixed and hybridized to a chip containing spots of DNA from genes 1 through 25. The results are shown in the adjoining illustration. What conclusions can you make about which genes might be implicated in antibiotic resistance in these bacteria? How might this information be used to design new antibiotics that are less vulnerable to resistance?

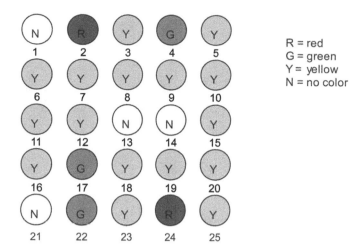

R = red
G = green
Y = yellow
N = no color

Solution:
Genes 2 and 24 are expressed at far higher levels in the antibiotic-resistant bacteria than in the nonresistant cells. Conversely, genes 4, 17, and 22 are downregulated. These genes may be involved in antibiotic resistance: upregulated genes may be involved in metabolism of the antibiotic or may perform functions that are inhibited by the antibiotic. Downregulated genes may be involved in import of the antibiotic or represent a cellular mechanism that accentuates the potency of the antibiotic. Characterization of these genes may lead to information regarding the mechanism of antibiotic resistance, and then to the design of new antibiotics that can circumvent this resistance mechanism.

19.   For the genes in the microarray illustrated in the lower part of **Figure 15.7**, are most overexpressed or underexpressed in tumors from patients that remained cancer free for at least five years? Explain your reasoning.

Solution:
Underexpressed. The expression of genes in tumors from patients that remained cancer free is shown as spots above the yellow line. Most of the spots are green, meaning the gene is underexpressed in tumor cells relative to noncancerous cells.

Section 15.3

20.   *Dictyostelium discoideum* is a soil-dwelling, social amoeba: much of the time, the organism consists of single, solitary cells, but, during times of starvation, individual amoebae come together to form aggregates that have many characteristics of multicellular organisms. Biologists have long debated whether *D. discoideum* is a unicellular or multicellular organism. In 2005, the genome of *D. discoideum* was completely sequenced. The table below lists some genomic characteristics of *D. discoideum* and other eukaryotes (L. Eichinger et al. 2005. *Nature* 435:43–57).

Feature	D. discoideum	P. falciparum	S. cerevisiae	A. thaliana	D. melanogaster	C. elegans	H. sapiens
Organism	Amoeba	Malaria parasite	Yeast	Plant	Fruit fly	Worm	Human
Cellularity	?	Uni	Uni	Multi	Multi	Multi	Multi
Genome size (millions bp)	34	23	13	125	180	103	2852
Number of genes	12,500	5268	5538	25498	13,676	19,893	22,287
Average gene length (BP)	1756	2534	1428	2036	1997	2991	27,000
Genes with introns (%)	69	54	5	79	38	5	85
Mean number of introns	1.9	2.6	1.0	5.4	4.0	5.0	8.1
Mean intron size (bp)	146	179	nd*	170	nd*	270	3365
Mean G + C (exons)	27%	24%	28%	28%	55%	42%	45%

*nd = not determined

a. On the basis of the organisms listed in the table other than *D. discoideum*, what are some differences in genome characteristics between unicellular and multicellular organisms?

Solution:
Unicellular organisms have smaller genomes and fewer genes.

b. On the basis of these data, do you think that the genome of *D. discoideum* is more like those of other unicellular eukaryotes or more like those of multicellular eukaryotes? Explain your answer.

Solution:
The genome size of *D. discoideum* is smaller, like unicellular organisms, but the number of genes approaches the number of genes in the *Drosophila* genome. Thus, the *D. discoideum* genome has characteristics of the genomes of both unicellular and multicellular organisms.

21. A group of 250 scientists sequenced and analyzed the genomes of 12 species of *Drosophila* (*Drosophila* 12 Genomes Consortium. 2007. *Nature* 450:203–218). Data on genome sizes and numbers of protein-encoding genes from this study are given in the accompanying table. Plot the number of protein-encoding genes as a function of genome size for 12 species of *Drosophila*. Is there a relation between genome size and number of genes in fruit flies? How does this result compare with the relation between genome size and number of genes across all eukaryotes?

Species	Genome size (millions of base pairs)	Number of protein-encoding genes
D. melanogaster	200	13,733
D. simulans	162	15,983
D. sechellia	171	16,884
D. yakuba	190	16,423
D. erecta	135	15,324
D. ananassae	217	15,276
D. pseudoobscura	193	16,363
D. persimilis	193	17,325
D. willistoni	222	15,816
D. virilis	364	14,680
D. mojavensis	130	14,849
D. grimshawi	231	15,270

Solution:
Any association between genome size and number of protein-encoding genes among the 12 *Drosophila* species is weak, at best (see below). This is consistent with the pattern seen among eukaryotic organisms in general, where there is no general association between genome size and number of genes.

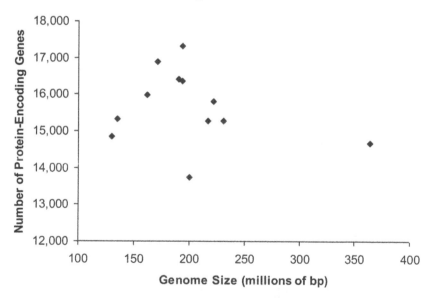

Section 15.4

*22. A scientist determines the complete genomes and proteomes of a liver cell and a muscle cell from the same person. Would you expect bigger differences in the genome or in the proteome of these two cell types? Explain your answer.

Solution:
Proteomes. Generally, the genomes of two cells from the same person will be genetically identical. However, the proteins of the two cells are likely to differ widely because different genes are expressed in each cell type.

CHALLENGE QUESTIONS

Section 15.1

23.   Some synthetic biologists have proposed the creation of an entirely new, free-living organism with a minimal genome—the smallest set of genes that allows for replication of the organism in a particular environment. This genome could be used to design and create, from "scratch," novel organisms that might perform specific tasks such as the breakdown of toxic materials in the environment.

    a. How might the minimal genome required for life be determined?

    Solution:
The minimal genome required might be determined by examining simple free-living organisms having small genomes to determine which genes they possess in common. Mutations can then be made systematically to determine which genes are essential for these organisms to survive. The apparently nonessential genes (those genes in which mutations do not affect the viability of the organism) can then be deleted one by one until only the essential genes are left. Elimination of any of these genes will result in loss of viability. Alternatively, essential genes could be assembled through genetic engineering, creating an entirely novel organism.

    b. What, if any, social and ethical concerns might be associated with the construction of an entirely new organism with a minimal genome?

    Solution:
The novel organism would prove that humans have acquired the ability to create a new species or form of life. Humans would then be able to direct evolution as never before. Among the social and ethical concerns would be whether human society has the wisdom to temper its power and whether such novel synthetic organisms can or will be used to develop pathogens for biological warfare or terrorism. After all, no person or animal would have been previously exposed or have acquired immunity to such a novel synthetic organism. There would be uncertainty about the new organism's effect on the ecosystem if it was released or escaped.

24.   The genome of the fruit fly *Drosophila melanogaster* was sequenced in 2000. However, this "completed" sequence did not include most heterochromatin regions. The heterochromatin was not sequenced until 2007 (Hoskins et al. 2007. *Science* 316:1625–1628). Most completed genome sequences do not include heterochromatin. Why is heterochromatin usually not sequenced in genomic projects? (Hint: See Chapter 8 for a more detailed discussion of heterochromatin.)

    Solution:
Heterochromatin is difficult to sequence because it contains many short repeated sequences. The presence of repeated sequences makes it difficult to assemble heterochromatin sequences into large contigs. This is analogous to trying to correctly assemble a jig-saw puzzle that contained many identical pieces. In addition, heterochromatin contains few genes, so geneticists have been less interested in heterochromatin sequences.

Chapter Sixteen: Cancer Genetics

COMPREHENSION QUESTIONS

Section 16.1

1. What types of evidence indicate that cancer arises from genetic changes?

   Solution:
   Many types of cancer are associated with exposure to radiation and other environmental mutagens. Some types of cancer tend to run in families, and a few cancers are linked to chromosomal abnormalities. Finally, the discovery of oncogenes and specific mutations that cause proto-oncogenes to become oncogenes or that inactivate tumor-suppressor genes proved that cancer has a genetic basis.

2. How can it be true that many types of cancer are genetic and yet not inherited?

   Solution:
   Most cancers are considered genetic diseases in the sense that they result from mutations in genes and chromosomes. Mutations in proto-oncogenes and tumor-suppressor genes, in particular, often disrupt regulation of the cell cycle and lead to cancer. Most commonly, these mutations occur in somatic cells and produce tumors in various tissues of the body. However, if the cancer-causing mutations occur in somatic cells and not in the germ line, the mutations will not be passed to the next generation. Sometimes mutations associated with cancer do occur in germ cells and are inherited. Even then, if the mutation is recessive, then a predisposition to cancer will be inherited rather than cancer itself.

3. Outline Knudson's two-hit hypothesis of retinoblastoma and describe how it helps to explain unilateral and bilateral cases of retinoblastoma.

   Solution:
   The multistage theory of cancer states that more than one mutation is required for most cancers to develop. Most retinoblastomas are unilateral because the likelihood of any cell acquiring two rare mutations is very low, and thus retinoblastomas develop in only one eye. Bilateral cases of retinoblastoma develop in people born with a predisposing mutation, and so only one additional mutational event will result in cancer. Thus, the probability of retinoblastoma development is higher in these persons and likely to be in both eyes. Because the predisposing mutation is inherited, people with bilateral retinoblastoma have relatives with retinoblastoma.

4. Briefly explain how cancer arises through clonal evolution.

   Solution:
   A mutation that relaxes growth control in a cell will cause it to divide and form a clone of cells that are growing or dividing more rapidly than their neighbors.

Successive mutations that cause even more rapid growth, or the ability to invade and spread, each produces progeny cells with more aggressive, malignant properties that outgrow their predecessors and take over the original clone.

Section 16.2

5.   What is the difference between an oncogene and a tumor-suppressor gene? Give some examples of functions of proto-oncogenes and tumor suppressors in normal cells.

Solution:
Oncogenes stimulate cell division, whereas tumor-suppressor genes put the brakes on cell growth. Proto-oncogenes are normal cellular genes that function in cell growth and in the *erbA*, *erbB*, *myc*, *src*, and *ras*. Tumor-suppressor genes inhibit cell-cycle progression; examples are *RB* and p53, which encode transcription factors.

6.   How do cyclins and CDKs differ? How do they interact in controlling the cell cycle?

Solution:
The CDKs, or cyclin-dependent kinases, have enzymatic activity and phosphorylate multiple substrate molecules when activated by binding the appropriate cyclin. Cyclins are regulators of CDKs and have no enzymatic activity of their own. Each cyclin molecule binds to a single CDK molecule. Whereas CDK levels remain relatively stable, cyclin levels oscillate in the course of the cell cycle.

7.   Briefly outline the events that control the progression of cells through the G_1/S checkpoint in the cell cycle.

Solution:
In G_1, cyclins D and E accumulate and bind to their respective CDKs. The cyclin D-CDK and cyclin E-CDK phosphorylate RB protein molecules. Phosphorylation of RB inactivates RB and releases active E2F protein. E2F protein transcribes genes required for DNA replication and progression into S phase.

8.   Why do mutations in genes that encode DNA-repair enzymes often produce a predisposition to cancer?

Solution:
Mutations that affect DNA repair cause high rates of mutation that may convert proto-oncogenes into oncogenes or inactivate tumor-suppressor genes. Similarly, errors in chromosome segregation cause aneuploidy and chromosomal aberrations that cause loss of tumor-suppressor genes or add extra gene doses of proto-oncogenes.

9.   What role do telomeres and telomerase play in cancer progression?

Solution:
DNA polymerases are unable to replicate the ends of linear DNA molecules. Therefore, the ends of eukaryotic chromosomes shorten with every round of DNA replication, unless telomerase uses its RNA component to add back telomeric DNA sequences, which it does in reproductive cells. Normally, somatic cells do not express telomerase; their telomeres progressively shorten with each cell division until vital genes are lost and the cells undergo apoptosis. Transformed cells (cancerous cells) induce the expression of the telomerase gene, thus leading to cell proliferation.

Section 16.3

10.   Briefly outline some of the genetic changes commonly associated with the progression of colorectal cancer.

Solution:
Colorectal cancer begins as benign tumors, called polyps, that enlarge and acquire further mutations that turn them malignant and, finally, invasive and metastatic. These progressive changes are associated with multiple mutations. One common sequence in colorectal cancer is a mutation of the APC gene that leads to faster cell division and polyp formation. Oncogenic mutations of the *ras* gene are found in cells from larger polyps. Mutations in p53 and other genes are found in malignant tumor cells, which may lead to genomic instability and additional changes that lead to greater malignancy and invasiveness.

Section 16.4

11.   Explain how chromosome deletions, inversions, and translocations can cause cancer.

Solution:
Deletions can cause the loss of one or more tumor-suppressor genes. Inversions and translocations can inactivate tumor-suppressor genes if the chromosomal breakpoints are within tumor-suppressor genes. Alternatively, a translocation can place a proto-oncogene in a new location, where it is activated by different regulatory sequences, causing the overexpression or unregulated expression of the proto-oncogene. Finally, inversions and translocations can bring parts of two different genes together, causing the synthesis of a novel protein that is oncogenic.

12.   Briefly outline how the Philadelphia chromosome leads to chronic myelogenous leukemia.

Solution:
The Philadelphia chromosome is a shortened chromosome 22 with a translocated tip of chromosome 9. A part of the c-*ABL* proto-oncogene from chromosome 9 is fused with *BCR* gene on chromosome 22. The resulting fusion protein is more active at promoting cell proliferation than the normal c-*ABL* protein, and causes leukemia.

13.   What is genomic instability? Give some ways in which genomic instability can arise.

Solution:
Genomic instability is a condition or process that leads to numerous chromosomal rearrangements and aneuploidy, often found in cells of advanced tumors. Mutations that affect the mitotic spindle checkpoint may cause a high frequency of aneuploidy. Other mutations, such as mutations in the *APC* gene, may affect the spindle itself or other aspects of the chromosome segregation mechanism. Still other mutations that affect centrosome duplication, such as some p53 mutations, could also lead to aneuploidy.

Section 16.5

14.   How do viruses contribute to cancer?

Solution:
Retroviruses have strong promoters. After its integration into a host genome, a retrovirus promoter can drive overexpression of a cellular proto-oncogene. Alternatively, the integration of a retrovirus can inactivate a tumor-suppressor gene. A few retroviruses carry oncogenes that are altered versions of host proto-oncogenes. Other viruses, such as human papilloma virus, produce proteins that affect the cell cycle.

APPLICATION QUESTIONS AND PROBLEMS

Introduction

15.   What characteristics of the pedigree shown in **Figure 16.1** suggest that pancreatic cancer in this family is inherited as an autosomal dominant trait?

Solution:
The presence of cancer and precancerous growths does not skip generations. Every affected individual has an affected parent. Pancreatic cancer and precancerous growths considered together are found in males and females.

Section 16.1

16.   If cancer is fundamentally a genetic disease, how might an environmental factor such as smoking cause cancer?

Solution:
Environmental factors can cause cancer by acting as mutagens. Higher rates of mutation will lead to higher rates of inactivation of tumor-suppressor genes or conversion of proto-oncogenes to oncogenes.

*17.   Both genes and environmental factors contribute to cancer. Prostate cancer is 30 times more common among people in Utah than among people in Shanghai (see **Table 16.2**). Briefly outline how you might go about determining if these differences in the incidence of prostate cancer are due to differences in the genetic makeup of two populations or differences in their environments.

Solution:
If the differences in cancer rates are due to genetic differences in the two populations, then people who migrated from Utah or Shanghai to other locations would have similar rates of cancer incidence as people who stayed in Utah or Shanghai. Moreover, different ethnic groups in Utah or Shanghai would have different rates of cancer. If the cancer rates are due to environmental factors, then people who migrated from Utah or Shanghai would have rates of cancer determined by their location and not by their place of origin, and different ethnic groups in the same location would have similar rates of cancer.

Section 16.2

*18.   The *palladin* gene, which plays a role in pancreatic cancer (see the introduction to this chapter), is said to be an oncogene. Which of its characteristics suggest that it is an oncogene rather than a tumor-suppressor gene?

Solution:
Because oncogenes promote cell proliferation, they act in a dominant manner. In contrast, mutations in tumor-suppressor genes cause loss of function and act in a recessive manner. When introduced into cells, the mutated *palladin* gene increases cell migration. Such a dominant effect suggests that *palladin* is an oncogene.

19.   Mutations in the *RB* gene are often associated with cancer. Explain how a mutation that results in a nonfunctional RB protein contributes to cancer.

Solution:
RB protein is a tumor suppressor, acting at the G_1/S checkpoint to prevent cells from beginning DNA replication. Without functional RB protein, cells are more prone to begin a round of cell division.

20.   Cells in a tumor contain mutated copies of a particular gene that promotes tumor growth. Gene therapy can be used to introduce a normal copy of this gene into the tumor cells. Would you expect this therapy to be effective if the mutated gene were an oncogene? A tumor-suppressor gene? Explain your reasoning.

Solution:
Gene therapy to introduce a normal copy of the gene into tumor cells will not work for oncogenes because oncogenes are dominant, activating mutations of proto-oncogenes. Gene therapy may work if the tumor arises from a mutation that inactivates a tumor-suppressor gene. Loss-of-function mutations are recessive; therefore, a normal copy of the

gene will be dominant and restore regulation of cell proliferation in the tumor cells. However, one would have to insert and express the tumor-suppressor gene in all tumor cells, which is not possible at this time.

21.   Some cancers have been treated with drugs that demethylate DNA. Explain how these drugs might work. Do you think the cancer-causing genes that respond to the demethylation are likely to be oncogenes or tumor-suppressor genes? Explain your reasoning.

Solution:
Drugs that demethylate DNA would presumably activate expression of demethylated genes. Cancer growth and progression may be inhibited if these drugs are able to turn on expression of tumor-suppressor genes that had been silenced by DNA methylation. If DNA demethylation turned on expression of oncogenes, cancer growth and progression would be accelerated.

Section 16.4

22.   Some cancers are consistently associated with the deletion of a particular part of a chromosome. Does the deleted region contain an oncogene or a tumor-suppressor gene? Explain.

Solution:
The deleted region contains a tumor-suppressor gene. Tumor suppressors act as brakes on cell proliferation. The deletion of tumor-suppressor genes will therefore permit the uncontrolled cell proliferation that is characteristic of cancer. Oncogenes, on the other hand, function as stimulators of cell division. Deletion of oncogenes will therefore prevent cell proliferation, and usually cannot cause cancer.

CHALLENGE QUESTIONS

Section 16.2

23.   Many cancer cells are immortal (will divide indefinitely) because they have mutations that allow telomerase to be expressed. How might this knowledge be used to design anticancer drugs?

Solution:
Because cancer cells depend on telomerase activity to preserve their telomeres, drugs that target telomerase enzymatic activity may limit the ability of cancer cells to divide indefinitely.

24.   Bloom syndrome is an autosomal recessive disease that exhibits haploinsufficiency. A recent survey showed that people heterozygous for mutations at the *BLM* locus are at increased risk of colon cancer. Suppose that you are a genetic counselor. A young woman is referred to you whose mother has Bloom syndrome; the young woman's father has no family history of Bloom syndrome. The young woman asks whether she is likely to experience any other health problems associated with her family history of Bloom syndrome. What advice would you give her?

Solution:
The young woman must be heterozygous for the mutation at the *BLM* locus because her mother was homozygous for the mutation. Although the young woman does not have Bloom syndrome, haploinsufficiency at this locus will result in some increased risk of colon cancer. Her cells will have a reduced amount of the *BLM* helicase involved in DNA double-strand break repair and will be more susceptible to mutations that may lead to cancer.

25.   Radiation is known to cause cancer, yet radiation is often used as treatment for some types of cancer. How can radiation be a contributor to both the cause and the treatment of cancer?

Solution:
Radiation can cause mutations that lead to cancer, such as inactivating a tumor-suppressor gene or causing an oncogenic mutation in a proto-oncogene. On the other hand, radiation will preferentially kill rapidly proliferating cells, such as cancer cells, that are actively replicating their DNA and lack tumor-suppressor functions that ensure DNA damage is repaired before DNA is replicated or before the cell divides.

26.   Imagine that you discover a large family in which bladder cancer is inherited as an autosomal dominant trait. Briefly outline a series of studies that you might conduct to identify the gene that causes bladder cancer in this family.

Solution:
Because this cancer is inherited as a dominant trait, the cause is most likely an oncogenic mutation, rather than inactivation of a tumor-suppressor gene. One possible approach to identify the bladder cancer oncogene would be to isolate DNA from bladder cancer cells and transfect cultured normal cells from a distinct genetic background. Isolate any colonies of transformed cells that arise and determine which common chromosomal region has been taken up by the transformed cell lines. The region of the gene could be further refined by transfection with subfragments of that region. Genes within this transforming region would be sequenced to identify mutations. Finally, candidate mutations would be tested by genetically engineering cells to carry that precise mutation to see if such cells become cancerous.

Chapter Seventeen: Quantitative Genetics

COMPREHENSION QUESTIONS

Section 17.1

1. How does a quantitative characteristic differ from a discontinuous characteristic?

   Solution:
   Discontinuous characteristics have only a few distinct phenotypes. In contrast, a quantitative characteristic shows a continuous variation in phenotype.

2. Briefly explain why the relation between genotype and phenotype is frequently complex for quantitative characteristics.

   Solution:
   Quantitative characteristics are polygenic, so many genotypes are possible. Moreover, most quantitative characteristics are also influenced by environmental factors. Therefore, the phenotype is determined by complex interactions of many possible genotypes and environmental factors.

3. Why do polygenic characteristics have many phenotypes?

   Solution:
   Many genotypes are possible with multiple genes. Even for the simplest two-allele loci, the number of possible genotypes is equal to 3^n, where n is the number of loci or genes. Thus, for three genes, we have 27 genotypes, four genes yield 81 genotypes, and so forth. If each genotype corresponds to a unique phenotype, then the number of phenotypes is the same: 27 for three genes and 81 for four genes. Moreover, the phenotype for a given genotype may be influenced by environmental factors, leading to an even greater array of phenotypes.

Section 17.2

4. What information do the mean and variance provide about a distribution?

   Solution:
   The mean is the center of the distribution. The variance is how broad the distribution is around the mean.

Section 17.3

5. List all the components that contribute to the phenotypic variance and define each component.

Solution:
V_G—component of variance due to variation in genotype
V_A—component of variance due to additive genetic variance
V_D—component of variance due to dominance genetic variance
V_I—component of variance due to genic interaction variance
V_E—component of variance due to environmental differences
V_{GE}—component of variance due to interaction between genes and environment

6. How do the broad-sense and narrow-sense heritabilities differ?

Solution:
Broad-sense heritability is the part of phenotypic variance due to all types of genetic variance, including additive, dominance, and genic interaction variances. Narrow-sense heritability is just that part of the phenotypic variance due to additive genetic variance.

7. Briefly describe common misunderstandings or misapplications of the concept of heritability.

Solution:
(1) Heritability is the portion of phenotypic variance due to genetic variance; it does not indicate to what extent the phenotype itself is determined by genotype.
(2) Heritability applies to populations; it does not apply to individuals.
(3) Heritability is determined for a particular population in a particular environment at a particular time. Heritability determined for one population does not apply to other populations, or even the same population facing different environmental conditions at a different period.
(4) A trait with high heritability may still be strongly influenced by environmental factors.
(5) High heritability does not mean that differences between populations are due to differences in genotype.

8. Briefly explain how genes affecting a polygenic characteristic are located with the use of QTL mapping.

Solution:
Two homozygous, highly inbred strains that differ at many loci are crossed and the F_1 are interbred. Quantitative traits are measured and correlated with the inheritance of molecular markers throughout the genome. The correlations are used to infer the presence of a linked QTL.

Section 17.4

9. How is the response to selection related to narrow-sense heritability and the selection differential? What information does the response to selection provide?

Solution:
The response to selection (R) = narrow-sense heritability (h^2) × selection differential (S). The value of R predicts how much the mean quantitative phenotype will change with different selection in a single generation.

10.   Why does the response to selection often level off after many generations of selection?

Solution:
After many generations, the response to selection plateaus because of two factors. First, the genetic variation may be depleted—all the individuals in the population now have the alleles that maximize the quantitative trait; with no genetic variation, there can be no selection or response to selection. Second, even if genetic variation persists, artificial selection may be limited by an opposing natural selection.

APPLICATION QUESTIONS AND PROBLEMS

Section 17.1

*11.   For each of the following characteristics, indicate whether it would be considered a discontinuous characteristic or a quantitative characteristic. Briefly justify your answer.

a.   Kernel color in a strain of wheat, in which two codominant alleles segregating at a single locus determine the color. Thus, there are three phenotypes present in this strain: white, light red, and medium red.

Solution:
Discontinuous characteristic because only a few distinct phenotypes are present and alleles at a single locus determine the characteristic.

b.   Body weight in a family of Labrador retrievers. An autosomal recessive allele that causes dwarfism is present in this family. Two phenotypes are recognized: dwarf (less than 13 kg) and normal (greater than 23 kg).

Solution:
Discontinuous characteristic because there are only two phenotypes (dwarf and normal) and a single locus determines characteristic.

c.   Presence or absence of leprosy. Susceptibility to leprosy is determined by multiple genes and numerous environmental factors.

Solution:
Quantitative characteristic because susceptibility is a continuous trait that is determined by multiple genes and environmental factors (an example of a quantitative phenotype with a threshold effect).

d. Number of toes in guinea pigs, which is influenced by genes at many loci.

Solution:
Quantitative characteristic because it is determined by many loci (an example of a meristic characteristic).

e. Number of fingers in humans. Extra (more than five) fingers are caused by the presence of an autosomal dominant allele.

Solution:
Discontinuous characteristic because only a few distinct phenotypes are determined by alleles at a single locus.

*12. Assume that plant weight is determined by a pair of alleles at each of two independently assorting loci (*A* and *a*, *B* and *b*) that are additive in their effects. Further assume that each allele represented by an uppercase letter contributes 4 g to weight and each allele represented by a lowercase letter contributes 1 g to weight.

a. If a plant with genotype *AA BB* is crossed with a plant with genotype *aa bb*, what weights are expected in the F_1 progeny?

Solution:
All weigh 10 g.

b. What is the distribution of weight expected in the F_2 progeny?

Solution:
We can group the 16 expected genotypes by the number of uppercase and lowercase alleles:
4 uppercase: *AA BB* = 1/16 with 16 g
3 uppercase: 2 *Aa BB*, 2 *AA Bb* – 4/16 with 13 g
2 uppercase: 4 *Aa Bb*, *aa* BB, *AA bb* = 6/16 with 10 g
1 uppercase: 2 *Aa bb*, 2 *aa Bb* = 4/16 with 7 g
0 uppercase: *aa bb* = 1/16 with 4 g

*13. Assume that three loci, each with two alleles (*A* and *a*, *B* and b, *C* and c), determine the differences in height between two homozygous strains of a plant. These genes are additive and equal in their effects on plant height. One strain (*aa bb cc*) is 10 cm in height. The other strain (*AA BB CC*) is 22 cm in height. The two strains are crossed, and the resulting F_1 are interbred to produce F_2 progeny. Give the phenotypes and the expected proportions of the F_2 progeny.

Solution:
The *AA BB CC* strain is 12 cm taller than the *aa bb cc* strain. We therefore calculate that each dominant allele adds 2 cm of height above the baseline 10 cm of the all-recessive strain. The F_1, with genotype *AaBbCc*, therefore will be 10 + 6 = 16 cm tall. The seven

different possible phenotypes with respect to plant height and the expected frequencies in the F_2 are listed in the following table:

Number of dominant alleles	Height (cm)	Proportion of F_2 progeny
6	22	1/64
5	20	6/64
4	18	15/64
3	16	20/64
2	14	15/64
1	12	6/64
0	10	1/64
		Total = 64/64

The proportions can be determined by counting the numbers of boxes with one dominant allele, two dominant alleles, and so on from an 8×8 Punnett square.

14.   Seed size in a plant is a polygenic characteristic. A grower crosses two pure-breeding varieties of the plant and measures seed size in the F_1 progeny. She then backcrosses the F_1 plants to one of the parental varieties and measures seed size in the backcross progeny. The grower finds that seed size in the backcross progeny has a higher variance than does seed size in the F_1 progeny. Explain why the backcross progeny are more variable.

Solution:
The F_1 progeny all have the same genetic makeup: they are all heterozygotes for the loci that differ between the two pure-breeding strains. The backcross progeny will have much greater genetic diversity as a result of the genetic diversity from meiosis of the F_1 heterozygotes.

Section 17.3

*15.   Phenotypic variation in the tail length of mice has the following components:
        Additive genetic variance (V_A)   = 0.5
        Dominance genetic variance (V_D)   = 0.3
        Gene interaction variance (V_I)   = 0.1
        Environmental variance (V_E)   = 0.4
        Genetic-environmental interaction variance (V_{GE}) = 0.0

    a. What is the narrow-sense heritability of tail length?

    Solution:
    0.38

b. What is the broad-sense heritability of tail length?

Solution:
0.69

16. The narrow-sense heritability of ear length in Reno rabbits is 0.4. The phenotypic variance (V_P) is 0.8 and the environmental variance (V_E) is 0.2. What is the additive genetic variance (V_A) for ear length in these rabbits?

Solution:
Narrow-sense heritability = V_A/V_P = 0.4
Given that V_P = 0.8, V_A = 0.4(0.8) = 0.32

17. A characteristic has a narrow-sense heritability of 0.6.

a. If the dominance variance (V_D) increases and all other variance components remain the same, what will happen to the narrow-sense heritability? Will it increase, decrease, or remain the same? Explain.

Solution:
The narrow-sense heritability will decrease. Narrow-sense heritability is V_A/V_P. Increasing the V_D will increase the total phenotypic variance V_P. If V_A remains unchanged, then the proportion V_A/V_P will become smaller.

b. What will happen to the broad-sense heritability? Explain.

Solution:
The broad-sense heritability V_G/V_P will increase. V_G is the sum of $V_A + V_D + V_I$. V_P is the sum of $V_A + V_D + V_I + V_E$. Increasing the numerator and denominator of the fraction by the same arithmetic increment will result in a larger fraction, if the fraction is smaller than 1, as must be the case for V_G/V_P.

c. If the environmental variance (V_E) increases and all other variance components remain the same, what will happen to the narrow-sense heritability? Explain.

Solution:
The narrow-sense heritability V_A/V_P will decrease because the total phenotypic variance V_P will increase if V_E increases.

d. What will happen to the broad-sense heritability? Explain.

Solution:
The broad-sense heritability V_G/V_P will decrease because $V_P = V_G + V_E$ will increase.

18.   What conclusion can you draw from **Figure 17.12** about the proportion of phenotypic variation in shell breadth that is due to genetic differences? Explain your reasoning.

Solution:
The narrow-sense heritability equals the proportion of the phenotypic variance that is due to additive genetic variance. Because the regression coefficient equals 0.7 and narrow-sense heritability equals the regression coefficient, the proportion of the phenotypic variance in shell breadth that is due to additive genetic variance is 0.7.

19.   Many researchers have estimated heritability of human traits by comparing the correlation coefficients of monozygotic and dizygotic twins. An assumption in using this method is that two monozygotic twins experience environments that are no more similar to each other than those experienced by two nonidentical twins. How might this assumption be violated? Give some specific examples of ways in which the environments of two identical twins might be more similar than the environments of two nonidentical twins.

Solution:
One obvious way monozygotic twins may have a more similar environment is if the dizygotic twins differ in sex. Dizygotic twins also differ more in physical traits than monozygotic twins. Such differences, in hair color, eye color, height, weight, and others, lead to different preferences in clothing, whether or when eye glasses or braces are required, and differences in preferred activities such as different aptitudes for sports.

*20.   A genetics researcher determines that the broad-sense heritability of height among Southwestern University undergraduate students is 0.90. Which of the following conclusions would be reasonable? Explain your answer.

  a. Because Sally is a Southwestern University undergraduate student, 10% of her height is determined by nongenetic factors.
  b. Ninety percent of variation in height among all undergraduate students in the United States is due to genetic differences.
  c. Ninety percent of the height of Southwestern University undergraduate students is determined by genes.
  d. Ten percent of the variation in height of Southwestern University undergraduate students is determined by variation in nongenetic factors.
  e. Because the heritability of height among Southwestern University students is so high, any change in the students' environment will have minimal effect on their height.

Solution:
Heritability is the proportion of total phenotypic variance that is due to genetic variance and applies only to the particular population. Thus, the only reasonable conclusion is **(d)**. Statement **(a)** is not justified because the heritability value does not apply to absolute height but to the variance in height among Southwestern undergraduates. Statement **(b)** is not justified because the heritability has been determined only for Southwestern University students; students at other universities, with different ethnic backgrounds and from different regions of the country may have different heritability for height. Statement **(c)** is

again not justified because the heritability refers to the variance in height rather than absolute height. Statement (e) is not justified because the heritability has been determined for the range of variation in nongenetic factors experienced by the population under study; environmental variation outside this range (such as severe malnutrition) may have profound effects on height.

21.   [Data Analysis Problem] *Drosophila buzzati* is a fruit fly that feeds on the rotting fruits of cacti in Australia. Timothy Prout and Stuart Barker calculated the heritabilities of body size, as measured by thorax length, for a natural population of *D. buzzati* raised in the wild and for a population of *D. buzzati* collected in the wild but raised in the laboratory (T. Prout and J. S. F. Barker. 1989. *Genetics* 123:803–813). They found the following heritabilities.

Population	Heritability of body size (± standard error)
Wild population	0.0595 ± 0.0123
Laboratory-reared population	0.3770 ± 0.0203

Why do you think that the heritability measured in the laboratory-reared population is higher than that measured in the natural population raised in the wild?

Solution:
Heritability is the proportion of total phenotypic variance that is due to genetic variance: $H^2 = V_G/V_P$. The difference in heritability between the wild population and the laboratory-reared population could be due to either the wild population having less genetic variance, or the wild population having greater phenotypic variance due to greater environmental or genetic–environmental interaction variance. Because the laboratory-reared population was collected in the wild, it is unlikely that the laboratory-reared population has greater genetic variance than the wild population. Therefore, the more likely explanation is that the wild population has greater phenotypic variance due to environmental factors. Variance due to differences in availability of food, parasitism, and ambient temperature may be some environmental factors.

22.   Mr. Jones is a pig farmer. For many years, he has fed his pigs the food left over from the local university cafeteria, which is known to be low in protein, deficient in vitamins, and downright untasty. However, the food is free and his pigs don't complain. One day, a salesman from a feed company visits Mr. Jones. The salesman claims that his company sells a new, high-protein, vitamin-enriched feed that enhances weight gain in pigs. Although the food is expensive, the salesman claims that the increased weight gain of the pigs will more than pay for the cost of the feed, increasing Mr. Jones' profit. Mr. Jones responds that he took a genetics class when he went to the university and that he has conducted some genetic experiments on his pigs; specifically, he has calculated the narrow-sense heritability of weight gain for his pigs and found it to be 0.98. Mr. Jones says that this heritability value indicates that 98% of the variance in weight gain among his pigs is determined by genetic differences, and therefore the new pig feed can have little effect on the growth of his pigs. He concludes that the feed would be a waste of his money. The salesman does not dispute Mr. Jones' heritability estimate, but he still claims that the new feed can significantly increase weight gain in Mr. Jones' pigs. Who is correct and why?

Solution:
The salesman is correct because Mr. Jones' determination of heritability was conducted for a population of pigs under one environmental condition: low nutrition. His findings do not apply to any other population or even to the same population under different environmental conditions. High heritability for a trait does not mean that environmental changes will have little effect.

Section 17.4

*23. Joe is breeding cockroaches in his dorm room. He finds that the average wing length in his population of cockroaches is 4 cm. He chooses six cockroaches that have the largest wings; the average wing length among these selected cockroaches is 10 cm. Joe interbreeds these selected cockroaches. From earlier studies, he knows that the narrow-sense heritability for wing length in his population of cockroaches is 0.6.

a. Calculate the selection differential and expected response to selection for wing length in these cockroaches.

Solution:
Use the equation: $R = h^2 \times S$, where S is the selection differential. In this case, $S = 10$ cm $- 4$ cm $= 6$ cm, and we are given that the narrow-sense heritability h^2 is 0.6. Therefore, the response to selection $R = 0.6(6$ cm$) = 3.6$ cm.

b. What should be the average wing length of the progeny of the selected cockroaches?

Solution:
The average wing length of the progeny should be the mean wing length of the population plus R: 4 cm + 3.6 cm = 7.6 cm.

24. Three characteristics in beef cattle—body weight, fat content, and tenderness—are measured, and the following variance components are estimated:

	Body weight	Fat content	Tenderness
V_A	22	45	12
V_D	10	25	5
V_I	3	8	2
V_E	42	64	8
V_{GE}	0	0	1

In this population, which characteristic would respond best to selection? Explain your reasoning.

Solution:
Tenderness would respond best because it has the highest narrow-sense heritability. The response to selection is given by the equation $R = h^2 \times S$, where the narrow-sense heritability h^2 is equal to V_A/V_P.

*25. A rancher determines that the average amount of wool produced by a sheep in her flock is 22 kg per year. In an attempt to increase the wool production of her flock, the rancher picks five male and five female sheep with the greatest wool production; the average amount of wool produced per sheep by those selected is 30 kg. She interbreeds these selected sheep and finds that the average wool production among the progeny of the selected sheep is 28 kg. What is the narrow-sense heritability for wool production among the sheep in the rancher's flock?

Solution:
0.75

26. A strawberry farmer determines that the average weight of individual strawberries produced by plants in his garden is 2 g. He selects the 10 plants that produce the largest strawberries; the average weight of strawberries among these selected plants is 6 g. He interbreeds these selected strawberry plants. The progeny of these selected plants produce strawberries that weigh 5 g. If the farmer were to select plants that produce an average strawberry weight of 4 g, what would be the predicted weight of strawberries produced by the progeny of these selected plants?

Solution:
Here we can use the equation $R = h^2 \times S$. R, the response to selection, is the difference between the mean of the starting population and the mean of the progeny of the selected parents. In this case, $R = 5$ g $- 2$ g $= 3$ g. S, the selection differential, is the difference between the mean of the starting population and the mean of the selected parents; in this case, $S = 6$ g $- 2$ g $= 4$ g. Substituting in the equation, we get 3 g $= h^2(4$ g$)$; $h^2 = 0.75$. If the selected plants averaged 4 g, then S would be 2 g and $R = 0.75(2$ g$) = 1.5$ g. Therefore, the predicted average weight of strawberries from the progeny plants would be 2 g + 1.5 g = 3.5 g.

27. Pigs have been domesticated from wild boars. Would you expect to find higher heritability for weight among domestic pigs or wild boars? Explain your answer.

Solution:
Wild boars will probably have higher heritability than domestic pigs. Domestic pigs, because of many generations of breeding and selection, are likely to have less variance and more homozygosity, for genes that affect commercial traits such as weight.

28. Has the response to selection leveled off in the strain of corn selected for high oil content shown in **Figure 17.14**? What does this observation suggest about genetic variation in the strain selected for high oil content?

Solution:
No. The percentage of oil content has continued to go up and shows no signs of leveling off. This suggests that genetic variation for oil content is still present in the strain.

CHALLENGE QUESTIONS

Section 17.1

29.   Bipolar illness is a psychiatric disorder that has a strong hereditary basis, but the exact mode of inheritance is not known. Research has shown that siblings of patients with bipolar illness are more likely also to develop the disorder than are siblings of unaffected individuals. Findings from one study demonstrated that the ratio of bipolar brothers to bipolar sisters is higher when the patient is male than when the patient is female. In other words, relatively more brothers of bipolar patients also have the disease when the patient is male than when the patient is female. What does this new observation suggest about the inheritance of bipolar illness?

Solution:
These observations suggest that an X-linked locus or loci may influence bipolar illness. Males inherit their X-chromosome genes only from their mother. Females inherit X-chromosome genes from both parents. Therefore, the brothers of an affected male inherited their X chromosome alleles from the same parent, the mother. On the other hand, an affected female may have inherited a contributory X-linked QTL locus allele from either parent; if this allele came from her father, there is no chance that her brothers inherited the same X-linked allele.

Section 17.3

30.   We have explored some of the difficulties in separating the genetic and environmental components of human behavioral characteristics. Considering these difficulties and what you know about calculating heritability, propose an experimental design for accurately measuring the heritability of musical ability.

Solution:
For the purpose of this discussion, let us assume that we have a reliable and accurate method of quantifying musical ability. I propose a study comparing musical abilities in individuals with different degrees of relatedness. I would compare two groups: one group would consist of monozygotic (identical) twins raised apart; the second group would consist of dizygotic (fraternal, or nonidentical) twins raised apart. Both groups should have comparable environmental variance, but the monozygotic twins share 100% of their genes, whereas the dizygotic twins share only 50% of their genes. By correlating the musical abilities of the two groups, we can estimate the broad-sense heritability with the following equation 22.20:

$$H^2 = 2(r_{MZ} - r_{DZ})$$

where r_{MZ} is the correlation coefficient of musical ability in the monozygotic group and r_{DZ} is the correlation coefficient in the dizygotic group.

Section 17.4

31.   [Data Analysis Problem] Eugene Eisen selected for increased 12-day litter weight (total weight of a litter of offspring 12 days after birth) in a population of mice (E. J. Eisen. 1972. *Genetics* 72:129–142). The 12-day litter weight of the population steadily increased but then leveled off after about 17 generations. At generation 17, Eisen took one family of mice from the selected population and reversed the selection procedure: in this group, he selected for *decreased* 12-day litter size. This group immediately responded to decreased selection; the 12-day litter weight dropped 4.8 g within one generation and dropped 7.3 g after five generations. On the basis of the results of the reverse selection, what is the most likely explanation for the leveling off of 12-day litter weight in the original population?

Solution:
The leveling off in the response to selection for increased litter weight may be due to either of two causes: elimination or reduction of genetic variance because maximum litter weight had been achieved, or opposing selection countering further increase in litter weight. The results of the reverse selection, showing immediate response, indicate that genetic variance was still significant. Therefore, the most likely explanation is opposing selection. Further increase in litter weight may cause problems for the mother during pregnancy, for example.

Chapter Eighteen: Population and Evolutionary Genetics

COMPREHENSION QUESTIONS

Section 18.1

1.   What is a Mendelian population? How is the gene pool of a Mendelian population usually described?

   Solution:
   A Mendelian population is a group of sexually reproducing individuals mating with each other and sharing a common gene pool. The gene pool is usually described by genotype frequencies and allele frequencies.

Section 18.2

2.   What are the predictions given by the Hardy–Weinberg law?

   Solution:
   The Hardy–Weinberg law states that a large population mating randomly with no effects from selection, migration, or mutation will have the following relationship between the genotype frequencies and allele frequencies:

   $f(AA) = p^2$; $f(Aa) = 2pq$; $f(aa) = q^2$, where p and q equal the allelic frequencies.

   Moreover, the allele frequencies do not change from generation to generation, as long as the above conditions hold.

3.   What assumptions must be met for a population to be in Hardy–Weinberg equilibrium?

   Solution:
   Large population, random mating, and not affected by migration, selection, or mutation

4.   Define inbreeding and briefly describe its effects on a population.

   Solution:
   Inbreeding is preferential mating between genetically related individuals. Inbreeding increases homozygosity and reduces heterozygosity in the population.

Section 18.3

5.   What determines the allelic frequencies at mutational equilibrium?

   Solution:
   At mutational equilibrium, the allelic frequencies are determined by the forward and reverse mutation rates.

6. What factors affect the magnitude of change in allelic frequencies due to migration?

    Solution:
    The proportion of the population due to migrants and the difference in allelic frequencies between the migrant population and the original resident population.

7. Define genetic drift and give three ways in which it can arise. What effect does genetic drift have on a population?

    Solution:
    Genetic drift is change in allelic frequencies resulting from sampling error. It may arise through a long-term limitation on population size, founder effect that occurs when the population is founded by a small number of individuals, or a bottleneck effect when the population undergoes a drastic reduction in population size. Genetic drift causes changes in allelic frequencies and loss of genetic variation because some alleles are lost as other alleles become fixed. It also causes genetic divergence between populations because the different populations undergo different changes in allelic frequencies and become fixed for different alleles.

8. What is effective population size? How does it affect the amount of genetic drift?

    Solution:
    The effective population size is the equivalent number of breeding adults in the population. The smaller the effective population size, the greater the magnitude of the genetic drift.

9. Define natural selection and fitness.

    Solution:
    Natural selection is the differential reproduction of genotypes. Fitness is the relative reproductive success of a genotype.

10. Briefly discuss the differences between directional selection, overdominance, and underdominance. Describe the effect of each type of selection on the allelic frequencies of a population.

    Solution:
    Directional selection occurs when one allele has greater fitness than another. Overdominance exists when the heterozygote has greater fitness than either of the homozygotes. Underdominance occurs when the heterozygote has less fitness than either of the homozygotes. Directional selection will cause the allele with greater fitness to increase in frequency and eventually reach fixation. Overdominance establishes a balanced equilibrium that maintains both alleles. Underdominance results in an unstable equilibrium that will degenerate once disturbed and move away from equilibrium until one allele is fixed.

11.   Compare and contrast the effects of mutation, migration, genetic drift, and natural selection on genetic variation within populations and on genetic divergence between populations.

Solution:
Mutation increases genetic variation within populations and increases divergence between populations because different mutations arise in each population.

Migration increases genetic variation within a population by introducing new alleles but it decreases divergence between populations.

Genetic drift decreases genetic variation within populations because it causes alleles to eventually become fixed but it increases divergence between populations because drift takes place differently in each population.

Natural selection either increases or decreases genetic variation, depending on whether the selection is directional or balanced. It either increases or decreases divergence between populations, depending on whether different populations have similar or different selection pressures.

Section 18.4

12.   What are the two steps in the process of evolution?

Solution:
Mutation and recombination

13.   How does anagenesis differ from cladogenesis?

Solution:
Anagenesis is change in a single group of organisms over time. Cladogenesis involves splitting of a group into two or more groups that become different from each other.

Section 18.5

14.   What is the biological species concept?

Solution:
The biological species concept defines species as a group of organisms whose members can potentially interbreed with one another but are reproductively isolated from members of other species.

15.   What is the difference between prezygotic and postzygotic reproductive isolating mechanisms?

Solution:
Prezygotic mechanisms operate before fertilization of the egg by sperm (or fusion of gametes), and postzygotic mechanisms operate after fertilization. Prezygotic mechanisms

include ecological isolation, behavioral isolation, temporal isolation, mechanical isolation, and gametic isolation. Postzygotic mechanisms include hybrid inviability, hybrid sterility, and hybrid breakdown.

16.   What is the basic difference between allopatric and sympatric modes of speciation?

Solution:
Allopatric speciation involves populations separated by a geographic barrier that precludes gene flow between the populations. Sympatric speciation takes place between populations occupying the same geographical area.

Section 18.6

17.   Briefly describe the difference between the distance approach and the parsimony approach to the reconstruction of phylogenetic trees.

Solution:
The distance approach relies on the degree of overall similarity between phenotypic characteristics or gene sequences; most-similar species are grouped together. The parsimony approach tries to reconstruct an evolutionary pathway on the basis of the minimum number of evolutionary changes that must have taken place since the species last had a common ancestor.

Section 18.7

18.   Outline the different rates of evolution that are typically seen in different parts of a protein-encoding gene. What might account for these differences?

Solution:
Nucleotide substitutions that do not change the amino acid sequence occur far more frequently than those that do change the amino acid sequence. Changes also occur more rapidly in those parts of the protein sequence that are not essential for the function of the protein. Nucleotide changes occur most rapidly in intron sequences, except for those positions that affect splicing. Selective pressure to maintain the function of proteins essential or advantageous for the survival and reproduction of the organism accounts for the differential rates of mutation and evolution.

19.   What is the molecular clock?

Solution:
The concept of a molecular clock is based on the idea that the rate at which nucleotide changes take place in a DNA sequence is relatively constant over long periods of time, and therefore the number of nucleotide substitutions that have taken place between two organisms can be used to estimate the time since they last shared a common ancestor.

20.   What is a multigene family? Which processes produce multigene families?

Solution:
Multigene families are a group of genes in the same genome that are related by descent from a common ancestral gene. Multigene families arise through gene duplication events with subsequent diversification. Some members acquire mutations that render them nonfunctional and become pseudogenes.

21.   Define horizontal gene transfer. What problems does it cause for evolutionary biologists?

Solution:
Horizontal gene transfer, also called lateral gene transfer, is transmission of genetic information across species boundaries. Horizontal gene transfer occurs frequently in bacteria, through transformation and phage-mediated transduction. In eukaryotes, horizontal gene transfer may occur through endosymbiosis (e.g., mitochondrial and chloroplast genes), viral infections, parasitic infections, and human intervention (genetic engineering). Horizontal gene transfers confound phylogenetic inference using molecular sequence data because the evolutionary history of horizontally transferred genes differs from the evolutionary history of the rest of the genome.

APPLICATION QUESTIONS AND PROBLEMS

Section 18.1

22.   How would you respond to someone who said that models are useless in studying population genetics because they represent oversimplifications of the real world?

Solution:
While it is important to keep in mind that models do represent simplifications, they nevertheless provide important and valuable information and predictions about the effects of size, mutation, migration, inbreeding, and selection on the gene pool of the population.

*23.   Voles (*Microtus ochrogaster*) were trapped in old fields in southern Indiana and were genotyped for a transferrin locus. The following numbers of genotypes were recorded, where T^E and T^F represent different alleles.

$T^E T^E$	$T^E T^F$	$T^F T^F$
407	170	17

Calculate the genotypic and allelic frequencies of the transferrin locus for this population.

Solution:
The total number of voles is 594.
$f(T^E T^E) = 407/594 = 0.685$
$f(T^E T^F) = 170/594 = 0.286$
$f(T^F T^F) = 17/594 = 0.029$

$$f(T^E) = f(T^E T^E) + f(T^E T^F)/2 = 0.685 + 0.143 = 0.828$$
$$f(T^F) = f(T^F T^F) + f(T^E T^F)/2 = 0.029 + 0.143 = 0.172$$

24.   [Data Analysis Problem] Jean Manning, Charles Kerfoot, and Edward Berger studied the allelic frequencies at the glucose phosphoglucose isomerase (GPI) locus in the cladoceran *Bosmina longirostris* (a small crustacean known as a water flea). They collected 176 animals from Union Bay in Seattle, Washington, and determined their GPI genotypes by using electrophoresis (J. Manning, W. C. Kerfoot, and E. M. Berger. 1978. *Evolution* 32:365–374).

Genotype	Number
$S^1 S^1$	4
$S^1 S^2$	38
$S^2 S^2$	134

Determine the genotypic and allelic frequencies for this population.

Solution:
$f(S^1 S^1) = 4/176 = 0.023$
$f(S^1 S^2) = 38/176 = 0.216$
$f(S^2 S^2) = 134/176 = 0.761$
$f(S^1) = (8 + 38)/352 = 0.13$
$f(S^2) = (268 + 38)/352 = 0.87$

Section 18.2

25.   A total of 6129 North American Caucasians were blood typed for the MN locus, which is determined by two codominant alleles, L^M and L^N. The following data were obtained:

Blood type	Number
M	1787
MN	3039
N	1303

Carry out a chi-square test to determine whether this population is in Hardy–Weinberg equilibrium at the MN locus.

Solution:
The total number of individuals is 6129.
$f(L^M) = p = (1787 + 3039/2)/6129 = 0.54$
$f(L^N) = q = (1303 + 3039/2)/6129 = 0.46$

A population in Hardy–Weinberg equilibrium should have the following genotype frequencies:
$f(L^M L^M) = p^2 = (0.54)^2 = 0.29 = 1777/6129$
$f(L^M L^N) = 2pq = 2(0.54)(0.46) = 0.50 = 3065/6129$
$f(L^N L^N) = q^2 = (0.46)^2 = 0.21 = 1287/6129$

Now we set up a chi-square test:

Blood type	Observed	Expected	O – E	(O – E)²	(O – E)²/E
M	1787	1777	10	100	0.056
MN	3039	3065	6	225	0.084
N	1303	1287	16	256	0.20

Chi-squared = Σ (O – E)²/E = 0.34

The number of degrees of freedom is the number of genotypes minus the number of alleles = 3 – 2 = 1.

Looking at a chi-square table, we see the *p* value is easily greater than 0.05, so we do not reject the hypothesis that this population is in Hardy–Weinberg equilibrium with respect to the MN locus.

26.   Assume that the phenotypes of lady beetles shown in **Figure 18.1b** are encoded by the following genotypes:

Phenotype	Genotype
All black	*BB*
Some black spots	*Bb*
No black spots	*bb*

a.   For the lady beetles shown in the figure, calculate the frequencies of the genotypes and frequencies of the alleles.

Solution:
f(*BB*) = 1/11 = 0.09
f(*Bb*) = 9/11 = 0.82
f(*bb*) = 1/11 = 0.09
f(*B*) = (2 + 9)/22 = 0.5
f(*b*) = (2 + 9)/22 = 0.5

b.   Use a chi-square test to determine if the lady beetles shown are in Hardy–Weinberg equilibrium.

Solution:
Expected genotypic frequencies under Hardy–Weinberg equilibrium:
$BB = p^2 \times N = (0.5)^2 \times 11 = 2.75$
$Bb = 2pq \times N = 2(0.5)(0.5) \times 11 = 5.5$
$Bb = q^2 \times N = (0.5)^2 \times 11 = 2.75$
chi-square = Σ (observed – expected)²/expected = ((1 – 2.75)²/2.75) + ((9 – 5.5)²/5.5) + ((1 – 2.75)²/2.75) = 1.11 + 2.23 + 1.11 = 4.45

Degrees of freedom = 3 – 2 = 1

Probability < 0.05

The probability that the difference between observed and expected is due to chance is less than 0.05 and therefore we conclude that the population is not in Hardy–Weinberg equilibrium.

27.   [Data Analysis Problem] Most black bears (*Ursus americanus*) are black or brown in color. However, occasional white bears of this species appear in some populations along the coast of British Columbia. Kermit Ritland and his colleagues determined that white coat color in these bears results from a recessive mutation (*G*) caused by a single nucleotide replacement in which guanine substitutes for adenine at the melanocortin-1 receptor locus (*mcr1*), the same locus responsible for red hair in humans (K. Ritland, C. Newton, and H. D. Marshall. 2001. *Current Biology* 11:1468–1472). The wild-type allele at this locus (*A*) encodes black or brown color. Ritland and his colleagues collected samples from bears on three islands and determined their genotypes at the *mcr1* locus.

Genotype	Number
AA	42
AG	24
GG	21

a.   What are the frequencies of the *A* and *G* alleles in these bears?

Solution:
$f(A) = (84 + 24)/174 = 0.62$
$f(G) = (42 + 24)/174 = 0.38$

b.   Give the genotypic frequencies expected if the population is in Hardy–Weinberg equilibrium.

Solution:
Expected genotype frequencies:
$f(AA) = (0.62)(0.62) = 0.384$
$f(AG) = 2(0.62)(0.38) = 0.471$
$f(GG) = (0.38)(0.38) = 0.144$

c.   Use a chi-square test to compare the number of observed genotypes with the number expected under Hardy–Weinberg equilibrium. Is this population in Hardy–Weinberg equilibrium? Explain your reasoning.

Solution:

Genotype	Observed	Expected	$O - E$	$(O - E)^2$	$(O - E)^2/E$
AA	42	33	9	81	2.45
AG	24	41	17	289	7.05
GG	21	13	8	64	4.92

Chi-squared $= \Sigma\,(O - E)^2/E = 14.42$

The number of degrees of freedom is the number of genotypes minus the number of alleles = $3 - 2 = 1$.

The p value is much less than 0.05; therefore, we reject the hypothesis that these genotype frequencies may be expected from Hardy–Weinberg equilibrium.

28. Genotypes of leopard frogs from a population in central Kansas were determined for a locus (M) that encodes the enzyme malate dehydrogenase. The following numbers of genotypes were observed:

Genotype	Number
$M^1 M^1$	20
$M^1 M^2$	45
$M^2 M^2$	42
$M^1 M^3$	4
$M^2 M^3$	8
$M^3 M^3$	6
Total	125

a. Calculate the genotypic and allelic frequencies for this population.

Solution:
$f(M^1 M^1) = 20/125 = 0.16$
$f(M^1 M^2) = 45/125 = 0.36$
$f(M^2 M^2) = 42/125 = 0.34$
$f(M^1 M^3) = 4/125 = 0.032$
$f(M^2 M^3) = 8/125 = 0.064$
$f(M^3 M^3) = 6/125 = 0.048$
$f(M^1) = p = 0.16 + 0.36/2 + 0.032/2 = 0.16 + 0.18 + 0.016 = 0.356$
$f(M^2) = q = 0.34 + 0.36/2 + 0.064/2 = 0.34 + 0.18 + 0.032 = 0.552$
$f(M^3) = r = 0.048 + 0.032/2 + 0.064/2 = 0.048 + 0.016 + 0.032 = 0.096$

b. What would the expected numbers of genotypes be if the population were in Hardy–Weinberg equilibrium?

Solution:
For population in Hardy–Weinberg equilibrium:
$f(M^1 M^1) = p^2 = (0.356)^2 = 0.127; 0.127(125) = 16$
$f(M^1 M^2) = 2pq = 2(0.356)(0.552) = 0.393; 0.393(125) = 49$
$f(M^2 M^2) = q^2 = (0.552)^2 = 0.305; 0.305(125) = 38$
$f(M^1 M^3) = 2pr = 2(0.356)(0.096) = 0.068; 0.068(125) = 8$
$f(M^2 M^3) = 2qr = 2(0.552)(0.096) = 0.106; 0.106(125) = 13$
$f(M^3 M^3) = r^2 = (0.096)^2 = 0.009; 0.009(125) = 1$

Genotype	Observed	Expected	$O - E$	$(O - E)^2$	$(O - E)^2/E$
M^1M^1	20	16	4	16	1
M^1M^2	45	49	−4	16	0.33
M^2M^2	42	38	4	16	0.42
M^1M^3	4	8	−4	16	2
M^2M^3	8	13	−5	25	1.9
M^3M^3	6	1	5	25	25

Chi-squared = 30.65
Degrees of freedom = # genotypes − # alleles = 6 − 3 = 3
The p value is much lower than 0.05; this population is not in Hardy–Weinberg equilibrium for this locus.

*29. Full color (D) in domestic cats is dominant over dilute color (d). Of 325 cats observed, 194 have full color and 131 have dilute color.

    a. If this population of cats is in Hardy–Weinberg equilibrium for the dilution locus, what is the frequency of the dilute allele?

    Solution:
    $f(\text{dilute}) = f(dd) = q^2 = 131/325 = 0.403$; $q = 0.635$

    b. How many of the 194 cats with full color are likely to be heterozygous?

    Solution:
    If $q = f(d) = 0.635$, then $p = 1 - q = 0.365$
    $f(Dd) = 2pq = 2(0.365)(0.635) = 0.464$; $0.464(325) = 151$ heterozygous cats

30. Tay–Sachs disease is an autosomal recessive disorder. Among Ashkenazi Jews, the frequency of Tay–Sachs disease is 1 in 3600. If the Ashkenazi population is mating randomly for the Tay–Sachs gene, what proportion of the population consists of heterozygous carriers of the Tay–Sachs allele?

Solution:
If q = the frequency of the Tay–Sachs allele, then $q^2 = 1/3600$; $q = 1/60 = 0.017$.
The frequency of the normal allele = $p = 1 - q = 0.983$.
The frequency of heterozygous carriers = $2pq = 2(0.983)(0.017) = 0.033$; approximately 1 in 30 are carriers.

*31. The human MN blood type is determined by two codominant alleles, L^M and L^N. The frequency of L^M in Eskimos on a small Arctic island is 0.80.

    a. If random mating takes place in this population, what are the expected frequencies of the M, MN, and N blood types on the island?

    Solution:
    0.64 for M, 0.32 for MN, and 0.04 for N.

b. If inbreeding is present in this population, what effect will it have on the expected numbers of the different blood types?

Solution:
With inbreeding, the frequencies of the M and N blood types will be higher than that expected with random mating, and the frequency of the MN blood type will be lower.

Section 18.3

*32. Pikas are small mammals that live at high elevation in the talus slopes of mountains. Most populations located on mountain tops in Colorado and Montana in North America are isolated from one another, because the pikas don't occupy the low-elevation habitats that separate the mountain tops and don't venture far from the talus slopes. Thus, there is little gene flow between populations. Furthermore, each population is small in size and was founded by a small number of pikas. A group of population geneticists proposes to study the amount of genetic variation in a series of pika populations and to compare the allelic frequencies in different populations. On the bases of the biology and the distribution of pikas, predict what the population geneticists will find concerning the within- and between-population genetic variation.

Solution:
The small population sizes and the founder effects would cause strong effects from genetic drift. The geneticists will find large variation between populations in allele frequencies. Within populations, the same factors coupled with inbreeding will cause loss of genetic variation and a high degree of homozygosity.

*33. Two chromosomal inversions are commonly found in populations of *Drosophila pseudoobscura:* Standard (*ST*) and Arrowhead (*AR*). When treated with the insecticide DDT, the genotypes for these inversions exhibit overdominance, with the following fitnesses:

Genotype	Fitness
ST/ST	0.47
ST/AR	1
AR/AR	0.62

What will the frequencies of *ST* and *AR* be after equilibrium has been reached?

Solution:
The selection coefficient is 1 − fitness, so the selection coefficients for the genotypes are

Genotype	Fitness	Selection coefficient
ST/ST	0.47	0.53
ST/AR	1.0	0.00
AR/AR	0.62	0.38

The chromosome inversions exhibit overdominance: the heterozygote has higher fitness than either homozygote. With overdominance, an equilibrium is reached, at which point the frequency of *AR* (*q*) will be

$$q = f(AR) = \frac{S_{AR/AR}}{S_{AR/AR} + S_{ST/ST}} = \frac{0.53}{0.53 + 0.38} = 0.58$$

The frequency of *ST* (*p*) will be $1 - q = 0.42$.

*34. [Data Analysis Problem] The fruit fly, *Drosophila melanogaster*, normally feeds on rotting fruit, which may ferment and contain high levels of alcohol. Douglas Cavener and Michael Clegg studied allelic frequencies at the locus for alcohol dehydrogenase (Adh) in experimental populations of *D. melanogaster* (D. R. Cavener and M. T. Clegg. 1981. *Evolution* 35:1–10). The experimental populations were established from wild-caught flies and were raised in cages in the laboratory. Two control populations (C1 and C2) were raised on a standard cornmeal–molasses–agar diet. Two ethanol populations (E1 and E2) were raised on a cornmeal–molasses–agar diet to which was added 10% ethanol. The four populations were periodically sampled to determine the allelic frequencies of two alleles at the alcohol dehydrogenase locus, *Adh*^S and *Adh*^F. The frequencies of these alleles in the experimental populations are shown in the graph.

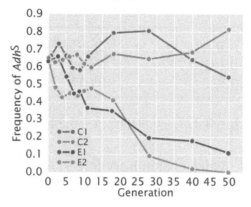

a. On the basis of these data, what conclusion might you draw about the evolutionary forces that are affecting the *Adh* alleles in these populations?

Solution:
The populations raised on the ethanol-containing diet appear to be experiencing directional selection in favor of the *Adh*^F allele, and against the *Adh*^S allele.

b. Cavener and Clegg measured the viability of the different *Adh* genotypes in the alcohol environment and obtained the following values:

Genotype	Relative viability
Adh^F/*Adh*^F	0.932
Adh^F/*Adh*^S	1.288
Adh^S/*Adh*^S	0.596

Using these relative viabilities, calculate relative fitnesses for the three genotypes.

Solution:
In the absence of data about relative reproductive rates, we use the relative viability data as a proxy for relative fitness.

$W_{FF} = 0.932/1.288 = 0.724$
$W_{FS} = 1.288/1.288 = 1.0$
$W_{SS} = 0.596/1.288 = 0.463$

The mean fitness $\bar{w} = p^2 W_{FF} + 2pq W_{FS} + q^2 W_{SS}$
$= 0.25(0.724) + 0.5(1.0) + 0.25(0.463) = 0.181 + 0.5 + 0.116 = 0.797$

We can then use the table method to calculate the frequencies after selection:

	$Adh^F Adh^F$	$Adh^F Adh^S$	$Adh^S Adh^S$
Initial genotypic frequencies:	$p^2 = (0.5)^2 = 0.25$	$2pq = 2(0.5)(0.5)$ $= 0.50$	$q^2 = (0.5)^2$ $= 0.25$
Fitnesses:	$W_{FF} = 0.724$	$W_{FS} = 1.0$	$W_{SS} = 0.463$
Proportionate contribution of genotypes to population:	$p^2 W_{FF} =$ $(0.25)(0.724)$ $= 0.181$	$2pq W_{FS} =$ $(0.50)(1.0)$ $= 0.50$	$q^2 W_{SS} =$ $(0.25)(0.463)$ $= 0.116$
Relative genotypic frequency after selection:	$p^2 W_{FF}/\bar{w} =$ $0.181/0.797 =$ 0.227	$2pq W_{FS}/\bar{w} =$ $0.50/0.797 =$ 0.627	$q^2 W_{SS}/\bar{w} =$ $0.116/0.797$ $= 0.145$

After one generation, $p = 0.227 + 0.5(0.627) = 0.54$.

35.   Examine **Figure 18.8**. Which evolutionary forces:

   a. cause an increase in genetic variation both within and between populations?

   Solution:
   Mutation, some types of natural selection

   b. cause a decrease in genetic variation both within and between populations?

   Solution:
   Some types of natural selection

   c. cause an increase in genetic variation within populations but cause a decrease in genetic variation between populations?

   Solution:
   Migration, some types of natural selection

Section 18.4

*36. The following illustrations represent two different patterns of evolution. Briefly discuss the differences in these two patterns with regard to how evolutionary change (on the *x* axis) occurs with respect to time (one the *y* axis).

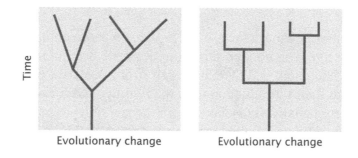

Time

Evolutionary change   Evolutionary change

Solution:
The two illustrations show the same pattern of cladogenesis. In each, an initial population splits into two groups with divergent evolutionary changes. Then, these two groups again each split into two other groups. The illustration on the left shows evolutionary differences accumulating gradually over time in each lineage after cladogenesis (diagonal lines indicate change accumulating over time); cladogenesis occurs before diversification and allows gradual diversification of the lineages. A population that is split by a geographic barrier may show this pattern. The illustration on the right shows splitting of a population into two apparently instantaneously divergent lineages; cladogenesis is accompanied by rapid diversification of the two lineages, which then remain stable for long periods of time until the next cladogenesis event (vertical lines indicate lack of change over time). This latter scenario may be produced by an evolutionary innovation in the population that instantly or very rapidly results in speciation.

Section 18.6

37. Which of the isolating mechanisms listed in **Table 18.3** have partly evolved between apple and hawthorn host races of *Rhagoletis pomonella*?

Solution:
Ecological (different host plants) and temporal (different times of mating)

38. [Data Analysis Problem] In Section 18.5, we considered the sympatric evolution of reproductive isolating mechanisms in host races of *Rhagoletis pomonella*, the apple maggot fly. The wasp *Diachasma alloeum* parasitizes apple maggot flies, laying its eggs on the larvae of the flies. Immature wasps hatch from the eggs and feed on the fly larvae. Research by Andrew Forbes and his colleagues (Forbes et al. 2009. *Science* 323:776–779) demonstrated that wasps that parasitize apple races of *R. pomonella* are genetically differentiated from those that parasitize hawthorn races of *R. pomonella*. They also found that wasps that prey on the apple race of the flies are attracted to odors from

apples, whereas wasps that prey on the hawthorn race are attracted to odors from hawthorns.

a. Propose an explanation for how genetic differences might have evolved between the wasps that parasitize the two races of *R. pomonella*.

Solution:
The apple and hawthorn races of *Rhagoletis pomonella* have developed some degree of reproductive isolation, because the flies mate near and lay their eggs on their respective host plant and there has been strong selective pressure for the flies to synchronize their reproduction with the timing of fruit ripening. Thus, flies that feed on apples are genetically isolated from those that feed on hawthorns. The wasps also appear to be attracted to the host plants; they probably use the odor of the fruits to find the flies that they parasitize. Because the wasps also mate near the host plants before laying their eggs on the flies, isolation of the host races of the flies will effectively keep the wasps that parasitize them reproductively isolated and allow wasps on different host races of *R. pomonella* to evolve genetic differences.

b. How might these differences lead to speciation in the wasps?

Solution:
Over time, genetic isolation between the wasps that parasitize the apple race of *R. pomonella* and those that parasitize the hawthorn race of *R. pomonella* might lead to postzygotic and prezygotic reproductive isolating mechanism and speciation.

39.   [Data Analysis Problem] Michael Bunce and his colleagues in England, Canada, and the United States extracted and sequenced mitochondrial DNA from fossils of Haast's eagle, a gigantic eagle that was driven to extinction 700 years ago when humans first arrived in New Zealand (M. Bunce et al. 2005. *Plos Biology* 3:44–46). Using mitochondrial DNA sequences from living eagles and those from Haast-eagle fossils, they created the phylogenetic tree below. On this phylogenetic tree, identify (a) all nodes; (b) one example of a branch; and (c) the outgroup.

Solution:
a. The internal nodes are all the branch points where lineages split.
b. The branches are the horizontal lines connecting nodes; one example of a branch is indicated with an arrow.
c. The outgroup is the Goshawk, the bottom branch and node in the figure.

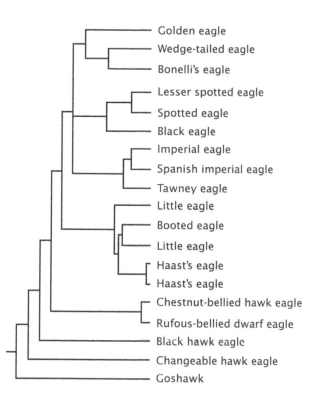

Golden eagle
Wedge-tailed eagle
Bonelli's eagle
Lesser spotted eagle
Spotted eagle
Black eagle
Imperial eagle
Spanish imperial eagle
Tawney eagle
Little eagle
Booted eagle
Little eagle
Haast's eagle
Haast's eagle
Chestnut-bellied hawk eagle
Rufous-bellied dwarf eagle
Black hawk eagle
Changeable hawk eagle
Goshawk

40. On the basis of the phylogeny of Darwin's finches shown in **Figure 18.12**, predict which two species in each of the following groups will be the most similar genetically.

   a. *Camarhynchus parvulus, Camarhynchus psittacula, Camarhynchus pallida*
   b. *Camarhynchus parvulus, Camarhynchus pallida, Platyspiza crassirostris*
   c. *Geospiza difficilis, Geospiza conirostris, Geospiza scandens*
   d. *Camarhynchus parvulus, Certhidea fusca, Pinaroloxias inornata*

   Solution:
   a. *Camarhynchus parvulus* and *Camarhynchus psittacula*
   b. *Camarhynchus parvulus* and *Camarhynchus pallida*
   c. *Geospiza conirostris* and *Geospiza scandens*
   d. *Camarhynchus parvulus* and *Certhidea fusca*

41. Assume that one of the genes shown in **Table 18.4** showed similar nonsynonymous and synonymous rates of substitutions. What might this suggest about the evolution of this gene?

   Solution:
   Similar rates of synonymous and nonsynonymous substitutions suggest that the gene does not encode a functional protein. For example, this is the pattern we often see in pseudogenes.

42. Based on the information provided in **Figure 18.17**, do introns or 3′ untranslated regions of a gene have higher rates of nucleotide substitution? Explain why.

Solution:
Introns. Higher rates of substitution are typically observed in those gene regions that have the least function, because natural selection limits variation in functional parts of genes. While the 3′ untranslated region of a gene does not encode amino acids, it does contain sequences that play a role in mRNA stability and translation. Within an intron, only sequences at the 5′ end, 3′ end, and branch point function in splicing.

CHALLENGE QUESTIONS

Section 18.3

43.   The Barton Springs salamander is an endangered species found in several springs at a single locality in the city of Austin, Texas. There is growing concern that a chemical spill on a nearby freeway could pollute the spring and wipe out the species. To provide a source of salamanders to repopulate the spring in the event of such a catastrophe, a proposal has been made to establish a captive-breeding population of the salamander in a local zoo. You are asked to provide a plan for the establishment of this captive-breeding population, with the goal of maintaining as much of the genetic variation of the species as possible in the captive population. What factors might cause loss of genetic variation in the establishment of the captive population? How could loss of such variation be prevented? With the assumption that only a limited number of salamanders can be maintained in captivity, what procedures should be instituted to ensure the long-term maintenance of as much of the variation as possible?

Solution:
Genetic variation in the zoo salamander colony could be reduced because of a founder effect from the limited number of individuals used to establish a breeding colony. Genetic variation would be reduced further by inbreeding and genetic drift. Given that only a limited number of salamanders can be maintained in the zoo colony, regular introduction of wild salamanders from the spring into the colony will keep mixing in fresh genotypes. A continual influx of migrants from the spring will, over time, effectively increase the number of individuals sampled, keep the gene pool of the zoo colony close to the gene pool of the spring, and mitigate inbreeding. It is also important to maintain a 50:50 sex ratio, as deviations from 50:50 causes a reduction in the effective number of adults. Matings within the zoo colony should be carefully planned to avoid inbreeding.

Section 18.5

44.   Explain why natural selection may cause prezygotic reproductive isolating mechanisms to evolve if postzygotic reproductive isolating mechanisms are already present, but natural selection can never cause the evolution of postzygotic reproductive isolating mechanisms.

Solution:

If postzygotic mechanisms are already present, then matings between individuals from the different groups result in no progeny. Such matings are wasted opportunities. Individuals with mutations that prevent such matings (prezygotic isolating mechanisms) have a higher fitness because all their matings are productive and will result in higher relative reproductive success compared to individuals that can waste some of their opportunities in unproductive matings. These mutations have a good chance of spreading through the population.

However, individuals with mutations that cause postzygotic reproductive isolation will have no progeny from some matings, whereas other individuals will have progeny from all their matings. Therefore, mutations that cause postzygotic reproductive isolation confer lower fitness and will tend to be eliminated from the population.